Britain's Military Helicopters

CHRIS GIBSON

MODERN MILITARY AIRCRAFT SERIES, VOLUME 4

Front cover image: AgustaWestland Wildcat HMA2 ZZ397 from 815 Naval Air Squadron on exercise in North West England. The Wildcat replaced the Lynx in Army Air Corps (AH1) and Fleet Air Arm (HMA2) service and will no doubt share the Lynx's longevity. (Peter Edwards)

Back cover image: Advancing forward to a landing in desert regions avoids a 'brown out' caused by dust thrown up by the rotor downdraught. There are few helicopters with a greater downdraught than the Chinook. This Chinook is delivering stores to 42 Commando, Royal Marines based at Garmsir in Afghanistan. (MOD/Open Government Licence)

Title page image: The first recognised use of aircraft to supply British forces by air occurred in 1915 when BE.2c biplanes dropped stores to a besieged garrison at Kut-al-Amara in Mesopotamia. Almost a century later, nothing much had changed: British aircraft were still delivering supplies to troops in the Middle East. A Chinook carries a pair of underslung 105mm light guns to gunners of 29 Commando Regiment, Royal Artillery in the desert of Oman. (MOD/Open Government Licence)

Contents page image: The Royal Marines operated the same Lynx models as the AAC with this AH7 fitted with wide skis for operations on snow for an exercise in northern Norway. The base for the operation was the amphibious assault ship HMS *Ocean*. (MOD/Open Government Licence)

Acknowledgements

It takes a team to produce a book such as *Britain's Military Helicopters,* but certain individuals deserve a special mention. Peter Edwards, Vic Flintham, Chris Lofting and James Lloyds kindly allowed use of their original photography, while the photo libraries of Phil Butler and Terry Panopalis always produce surprises. James Jackson read the draft and clarified a number of points, especially on matters naval, while Hugh Lake proved invaluable on the Air Staff's Chinook procurement process. Jessica Brown edited the text and clarified a number of points related to helicopter training operations. This book would not have been possible without the insight and guidance of former Westland employees Jeremy Graham and Ron Smith. Thanks are also due to the crew of the *Ocean Patriot,* particularly Stuart Ruddiman and Rob Douglas.

Published by Key Books
An imprint of Key Publishing Ltd
PO Box 100
Stamford
Lincs PE19 1XQ

www.keypublishing.com

The right of Chris Gibson to be identified as the author of this book has been asserted in accordance with the Copyright, Designs and Patents Act 1988 Sections 77 and 78.

Copyright © Chris Gibson, 2021

ISBN 978 1 80282 026 3

All rights reserved. Reproduction in whole or in part in any form whatsoever or by any means is strictly prohibited without the prior permission of the Publisher.

Typeset by SJmagic DESIGN SERVICES, India.

Contents

Introduction 4
Chapter 1 Where Nothing Else Can Go 5
Chapter 2 What Are Helicopters For? 7
Chapter 3 A Practical Use 11
Chapter 4 Meeting the Navy's Needs 14
Chapter 5 The Army Takes to the Air 33
Chapter 6 Assault and Support by the RAF 36
Chapter 7 The Army and Its Helicopters 59
Chapter 8 Rotary Training 73
Chapter 9 The Helicopter in Royal Service 79
Chapter 10 For Those in Peril 84
Conclusion 92
Appendices 93
Select Bibliography 95
Glossary 96

Introduction

The immediately recognisable sound of a 'chopper' can signify many things: rescue for those in peril on the sea, impending doom for a submarine under the waves, medical evacuation for wounded infantry and death from above for an enemy. For this author, the clatter of a helicopter means going home after three weeks at sea. Despite having none of their own, Britain's armed forces embraced the helicopter from its earliest days in World War Two, whether it be used for transferring kit between ships underway or evacuating wounded soldiers in Burma. From these early experiences with American-operated Sikorsky R-4Bs, Britain's forces soon acquired helicopters of their own.

While Fleet Air Arm officers had inspected and even flown Sikorsky VS-300s, leading to the Navy conducting trials, the RAF's first helicopters, the Sikorsky R-4B, designated as the Hoverfly I, arrived in February 1945 at the RAF's Helicopter Training Flight at Andover, and aside from ongoing arguments between the Army and the Air Force over which arm should operate which type in whatever role, the helicopter has thrived in British service. Of course, the Royal Navy had a much more focused approach to the helicopter and steadfastly avoided the Army-Air Force turf war.

In *Britain's Military Helicopters*, the various types and roles of the helicopter in the service of Britain's armed forces will be examined and some quite extraordinary background uncovered, such as how a British equivalent of the US Army's Air Cavalry could have been operating in 1944 and how the Chinook entered service thanks to a piece of crafty artwork and imaginative accounting.

Almost 80 years after their introduction to military service in the United Kingdom, the Army-Air Force turf war continues, while the Navy watches from afar and 'just gets on with it'.

Above: The original helicopter for the Royal Navy. A Sikorsky R-4 Hoverfly I takes to the air at Old Rockaway Airport, New York. The Navy knew what it wanted from the helicopter, but it would be decades before the Navy's foresight became reality. (USN via Terry Panopalis)

Left: Igor Sikorsky built and flew the VS-300, the first single-rotor helicopter, in September 1939, and very soon, Royal Navy pilots had flown it. The VS-300 would lead to the R-4B and a lineage of Sikorsky machines. (Blue Envoy Collection)

Chapter 1
Where Nothing Else Can Go

Be it in logistics, air-assault, air-support or anti-submarine roles, the helicopter has become a key element in modern warfare. Back in the late 1940s, when the helicopter appeared on the scene, the armed services were a bit confused by what they could do and who should operate them. The Hoverfly I had a payload on a par with the Auster AOP7 (Air Observation Post) and was more complicated to fly and operate, with its only benefit being vertical take-off and landing. That would prove to be the helicopter's trump card – it could go where no other aircraft could and do things no other aircraft could.

Not that the rotary winged aircraft was absent from the British air forces, the RAF operated Avro-built Cierva C.30 Rota I autogyros in the radar calibration and Army Co-operation roles throughout the war. However, their limited payload and need for a rolling take-off (unless the 'jump start' was employed) made them less than practical for the RAF. The Royal Navy acquired a pair of Cierva C.40s for evaluation, but, like the RAF, found them wanting and soon turned its attention to the helicopter being developed by Igor Sikorsky.

While the armed forces were trying to work out what to do with the small helicopters available, the British aircraft companies were doing the government's bidding. Immediately after World War Two, a company

Right: The RAF had long experience with autogyros (or Autogiro if referring to Cierva machines), and these were used for radar calibration during the war. This Avro Rota HM580 is a licence-built Cierva C.40. (Blue Envoy Collection)

Below: The first helicopter in British service was the Sikorsky R-4B, designated Hoverfly I by the RAF, with KK995 being one of the first to arrive from the USA in November 1944. KK995 was one of the Hoverflies used to train the AOP pilots of 657 Sqn. (Blue Envoy Collection)

called Pest Control Ltd wanted a machine to use as a crop sprayer. This was to meet Civil Aircraft Operational Requirement CAOR 3/46 and resulted in a utilitarian three-rotor machine, the Cierva W.10, nicknamed the 'Spraying Mantis'. The W.10 was a testbed/prototype for a larger machine, the Cierva W.11, also known as the Air Horse, and at the time of its first flight in December 1948, it was the largest helicopter in the world. Powered by a single Rolls-Royce Merlin engine, the three-rotor Air Horse had a useful payload of 6,720lb (3,050kg) i.e. 3 tons, the same as a standard army truck. This attracted the Army's attention and, despite being a civil aircraft and a testbed to boot, it showed that large helicopters were viable. Plans were afoot to increase the capability of the machine by adding another Merlin or powering an enlarged Air Horse with a pair of Armstrong Siddeley Mamba turboshafts, designated as the Cierva W.11T.

The Air Horse's payload of 3 tons caught the attention of the Army as an alternative way to meet Specification X30/46 for an assault glider to replace the Airspeed Horsa, which incidentally had a smaller payload than the Air Horse at 5,821lb (2,640kg). Oddly enough, the General Staff viewed the Air Horse as a tug for gliders rather than an assault machine in its own right.

Meanwhile at the Air Ministry, the Air Staff appeared ambivalent towards the helicopter. The latest Hoverfly II (Sikorsky R-6) equipped the Helicopter Training Flight (HTF) where pilots were trained, but the HTF also undertook trials to assess the practical application of the machines. These turned out to be little more than the previous Hoverfly I, so aside from training, continuation flying and delivering the King's mail, the RAF gave the impression they had little interest in doing much with helicopters. This situation was mainly owing to the lack of a helicopter that could be used for a practical military purpose.

The Royal Navy and the British Army never suffered from the RAF's ambivalence. The senior services had multiple applications for helicopters: anti-submarine, search and rescue and plane guard for the Navy, while the Army could use them for airborne observation post (AOP), reconnaissance, liaison and resupply. For the RAF, its ambivalence was short-lived, and the transformation from a machine with little practical use into a capable necessity came in Malaya during 1950.

The real game changer for the helicopter, enabling its evolution from an observation and liaison type into a practical military machine, was the gas turbine. Light and powerful, the turboshaft is a gas turbine whose output is delivered via a shaft, making it a compact and powerful alternative to the heavy and maintenance-intensive reciprocating engine and their associated ground resonance problems. Since the early 1960s, the only helicopters with piston engines have been in the lightweight class; effectively, helicopters as they were in the 1940s.

With the introduction of the turboshaft, helicopters went from strength to strength, and by the 1970s they had become key elements of any armed force's order of battle. Whether it was hunting and destroying submarines, knocking out tanks or deploying troops and equipment where and when required, the gas-turbine helicopter created new roles and added to the flexibility and capability of all the armed services in the United Kingdom.

The Cierva W.11 Air Horse was, at the time of its first flight in 1948, the largest and most powerful helicopter in the world. Considered for a variety of roles including glider tug, the Air Horse was to form the basis of the British Army's Hover Force. (Blue Envoy Collection)

Chapter 2
What Are Helicopters For?

As noted in the Introduction, the three British armed services differed in their assessment of the practicality of the helicopter. While the Royal Navy knew exactly what they wanted, an anti-submarine warfare (ASW) platform, and the British Army had a long list of roles, the RAF was vague on applications.

A Royal Navy mission in the USA had, in 1942, evaluated the Sikorsky VS-300, the first successful single-rotor helicopter and the blueprint for subsequent designs. The mission members soon got their hands on the prototype R-4B and became very interested in the US Navy's work in fitting a dipping sonar to the R-4B. Unfortunately, the R-4B was unsuitable for ASW work but Sikorsky's S-51, designated R-5, was welcomed by the Admiralty, with two being ordered for evaluation. The R-5 would be licence-built by Westland as the WS-51 Dragonfly, but rather than use it for ASW, the Royal Navy used it for search and rescue, plane guard and transport.

Neither the later Sikorsky S-55, its Westland-built successors nor the projected Bristol Type 191 were able to meet the Admiralty's ASW needs, effectively hobbled by their reciprocating powerplants. It would take two, if not three, generations of helicopter to produce the true 'Single Package' ASW, with dipping sonar and torpedo on the same machine, that the Admiralty sought.

The Westland WS-51 Dragonfly became a stalwart of the Fleet Air Arm's helicopter units but offered restricted capacity in whatever role they filled. (via James Jackson)

The post-war Army's list of roles for the helicopter could almost have been the same list that Victorian armies had for the horse: transport, reconnaissance, communication and exploitation of breakthroughs. This last item, essentially using helicopters to deploy troops and weapons, dated back to a 1940 letter written by Lt Col LVS Blacker, one of the War Office's most talented weapons developers, who suggested using a 'Rotaplane' to deploy troops. Two years later, what could be described as the first attack helicopter, the Hafner PD.7 armed with a 20mm Hispano cannon, was shown to the War Office and Air Ministry. Raoul Hafner's design may be the 'slow flying triplane' discussed in a General Staff paper on close support aircraft.

In 1948, the Director of Land/Air Warfare proposed 'Hover Force': a large rapid reaction force that could rapidly deploy troops and firepower to areas where the enemy had broken through. Hover Force would include reconnaissance machines, troop carriers, anti-tank platforms and defence suppression helicopters, all based on the only large helicopter available, Cierva's W.11 Air Horse. Sadly, the sound concept that was the Hover Force would not see the light of day until 1965 when the 1st Cavalry Division of the US Army deployed a similar formation in the central highlands of South Vietnam.

The Hover Force, as laid out in the Director of Land/Air Warfare's 1948 paper, has provided the blueprint for the use of helicopters in a ground war or, if the forward base is a ship, amphibious assaults. The problem facing a putative Hover Force was the same as that stalling development of the Navy's ASW platform – lack of lifting capacity and the need for large and heavy helicopters to fulfil the requirements. Fortunately, the solution to the problem was readily available in the gas turbine and the switch from piston to gas-turbine engines on the Admiralty's helicopters made the 'Single Package' ASW platform viable and enabled the Army to circumvent the Air Staff's attempts to restrict the use of battlefield helicopters.

The Turbine Era

Consider the powerplant on the Westland WS-55 Whirlwind. The Whirlwind was powered by the Alvis Leonides Major 755 piston engine and, much to the chagrin of the Admiralty and Air Staff, was severely limited in capability. The conversion of HAS7s to HAR9s with gas-turbine power by installing the de Havilland Gnome turboshaft transformed the machine's performance not only through increased power, but also reduced powerplant weight.

Westland Sikorsky WS-55 Whirlwind I G-OACZ was the Westland demonstrator for the early Westland-built S-55s powered by the Pratt & Whitney R-1340. (Blue Envoy Collection)

What Are Helicopters For?

Name	Rating	Dry Weight	Power/Wt. Ratio	Diameter	Length
Alvis Leonides Major 755	750hp (560kW)	1,200lb (544kg)	0.625	39in (99cm)	71in (180cm)
de Havilland Gnome H-1000	1,050shp (783kW)	334lb (152kg)	3.14	23in (58cm)	55in (140cm)

While the benefits of using turboshafts to power helicopters had been clear from the start, there were doubts about the reliability of engines. By the early 1950s, the gas turbine was established as a powerplant in fixed-wing aircraft and the gas turbine soon appeared on helicopters. Another reason for the adoption of the turboshaft was the Royal Navy's drive to rid its ships of gasoline, and as the largest customer for Westland's wares, the gas turbine made sense. The Sikorsky S-58, for which Westland had a licence to build, was powered by a 1,525hp (1,137kW) Wright Cyclone R-1820-84 with a dry weight of 1,184lb (537kg). For the Wessex HAS1, Westland replaced the Cyclone with a Napier Gazelle rated at 1,450shp (1,081kW) and with a dry weight of 830lb (377kg). Although the Gazelle rating was lower than the Cyclone, the weight difference was significant. The later Wessex HU5 and HC2 were powered by two de Havilland Gnome H.1200 turboshafts rated at 1,550shp (1,156kW) and a dry weight of around 670lb (304kg).

Besides increased power, the gas turbine revolutionised the world of helicopter design. While conversions of piston types such as the Whirlwind transformed their performance, new designs tailored to the turboshaft were now possible with the roof-mounted powerplant becoming the norm.

By fitting the Rolls-Royce Gnome, Westland radically improved the Whirlwind's performance. Whirlwind G-APDY was Westland's demonstrator for the Gnome Whirlwind. (Terry Panopalis Collection)

The turboshaft could now be installed above the cabin with the cockpit in front and fuel below, producing a much more compact 'box' configuration with more usable volume in the fuselage.

The Wessex was the first British production helicopter powered by a gas turbine, and the first Gazelle-powered Wessex HAS1 was delivered in 1960. However, it was the updated HAS3 that provided the Fleet Air Arm with a most effective ASW platform, although not strictly the single-package ASW type it had sought since the late 1940s. The other Westland type converted to gas-turbine power was the Whirlwind, and the substitution of the Leonides Major for the Gnome in 1960 transformed the type. A disappointment in Malaya, the Whirlwind became a mainstay of the Royal Navy (RN), and particularly the RAF, helicopter forces around the world until 1982.

Westland took the piston-engined Sikorsky S-58 and produced the Wessex, with XL722 as demonstrator. Initially powered by a single Gazelle on the HAS1 and HAS3, the later Coupled Gnome for the HC2 and HU5 offered twin-engined safety over the sea. (Blue Envoy Collection)

Chapter 3

A Practical Use

Having evaluated the Sikorsky R-4B Hoverfly I and R-6 Hoverfly II for a couple of years, the RAF had concluded that their development potential was exhausted and they had little operational use. Many of the R-4Bs had been passed to the Navy and the R-6 was being used by the Army for AOP development. Unlike these services, the RAF could see little use for the helicopter and the Royal Navy, in particular, more or less monopolised helicopter requirements into the late 1950s. Another aspect that affected the RAF's helicopter requirements was the drive by British helicopter companies to develop machines aimed at the civil market, with the helicopter viewed as the future of short-range air transport. This was the role that the British government put its money on in the late 1940s and early 1950s. Aside from supporting offshore oil operations, commercial passenger transport helicopters have failed to appear. This view quite probably led to the ascendance of French companies in the field of military helicopters in Europe and a growing British dependence on licence-building US types.

Meanwhile in Malaya, communist terrorists had attacked British-owned estates in June 1948, prompting the declaration of a state of emergency in the Malay Peninsula. The Malayan Emergency lasted from 1948 until 1960 and the RAF's involvement was designated Operation *Firedog*. By 1950, the counter-insurgency war saw operations deep inside the Malayan jungle, which made casualty

The WS-51 could be fitted with two litters for the medevac role, but their weight plus two casualties was beyond the machine's capability in the hot and high conditions of Malaya. The litters were dispensed with and a single lightweight litter for internal carriage was developed. (Blue Envoy Collection)

evacuation by fixed-wing aircraft, such as the Auster AOP5, more or less impossible, as they needed an airstrip of some description.

In late 1949, the RAF was called on to provide casualty evacuation (casevac) for ground troops, and the Far East Air Force (FEAF) established the Casualty Evacuation Flight (CEF) on 1 May 1950 at Seletar in Singapore. Lacking a suitable helicopter, the flight operated a trio of Westland-Sikorsky WS-51 Dragonfly HC2s transferred from the Royal Navy and conducted its first casualty evacuation on 14 June 1950. As operations continued, the flight developed equipment and techniques for the role and, defying the sceptics, the Dragonfly proved its worth. The unit traded its Dragonfly HC2s for HC4s (derived from the Navy's HR3) in 1952 and the CEF formed the basis of 194 Sqn in February 1953.

The Dragonfly initially operated with a casualty litter on each side of the cabin, but operations in the hot and high conditions in Malaya soon revealed that the machine could only operate with a single pilot if two litters were fitted. With a single litter, a second crewman could be carried to load the litter and guide the pilot when landing in forest clearings. Being external, the casualty could not be attended to in flight and the heavy external litters were soon removed and a bespoke, locally made litter that could be accommodated within the cabin was adopted. Internal carriage of the casualty also helped the operation of the Dragonfly, as it was sensitive to centre of gravity changes that required ballast to be moved around the airframe depending on the loading.

Seeking a more capable type for operations in Malaya, including demands from the Army for troop transports, the Air Staff asked for Whirlwinds. It soon became apparent that the Royal Navy had ordered all the available Westland Whirlwinds and none were available for the RAF! Ordering Sikorsky S-55s from the US under the Mutual Defense Aid Program (MDAP) was not possible because the Malayan Emergency was deemed a colonial operation and equipment supplied under the MDAP could only be used for NATO operations. Of course, there were ways around such obstacles, and the upshot was that the Royal Navy provided a squadron, 848 Naval Air Squadron (NAS), equipped with US-built S-55s. These were replaced by the Westland-built Whirlwind HAR1 and HAR4 as soon as the Westland machines became available in early 1954.

The Whirlwind was a much more practical proposition thanks to its large cabin under the main rotor, and loads, including troops, could be airlifted without the centre of gravity considerations of the Dragonfly. Capable of carrying up to four troops, the Whirlwind enabled the Army, RAF and RN to develop the techniques for heliborne assault that would become the 'bread and butter' of future helicopter operations.

The Westland-built Whirlwinds differed in many ways from the S-55, notably in airframe weight, which was calculated as being 5% higher than the US-built models thanks to heavier gauge sheet metal used in its construction. Another problem was 40 gallons (182 litres) of fuel in rear tanks that could not be used because of changes in the fuel system of the Westland machines! Unfortunately, some of the modifications to address the higher weight, specifically those to the engine supercharger, backfired and the initial Whirlwind HAR2s could not match the load capacity of the smaller types such as the Bristol Sycamore. The RAF's Engineering Branch worked tirelessly on the problems and ultimately solved them. Despite these disadvantages, the Whirlwinds were required to replace the S-55s not only for political reasons: they were also being heavily used. Not that the RAF was alone in its Whirlwind woe: the Royal Navy was finding similar problems with its HAR1 fleet.

By 1953, the first British-designed helicopter arrived in RAF service in the shape of the Bristol Type 171 Sycamore. When the CEF had been proposed in late 1949, the Type 171 had been considered but dismissed in favour of the Dragonfly. There were several reasons for this, including the low height of the Sycamore's rotor disc, but the main factor was that the Type 171 was not yet available. By 1952, the Type 171 Sycamore was a known quantity and promised a much-improved performance in the casevac role. This was borne out in operational trials of an HC10 in Malaya during early 1953 when, three days into the evaluation period and with no Dragonflies available, the trials Sycamore was called upon to evacuate a wounded soldier. On completion of the trials, the Sycamore HR14 entered service in Malaya, replacing the Dragonfly from late 1954 in the casevac role. The bulged passenger doors of the Sycamore enabled carriage of two stretchers, one above the other, but this was rarely done. Where the Sycamore really scored over the Dragonfly was serviceability and availability, and they operated successfully until 1960 before being transferred to Brunei.

The Sycamore served well in Malaya, and this example, HR14 XJ918, has an interesting history. It served with the first RAF SAR squadron, 275 at Thornaby before moving to Leconfield. It was then lent to the Aircraft and Armament Experimental Establishment (A&AEE), who used it to investigate rotor problems in Malaya, before being transferred to Brunei. After a period as the personal transport of the AOC FEAF, it was returned to the UK and now resides in the RAF Museum Cosford. (Blue Envoy Collection)

Chapter 4
Meeting the Navy's Needs

The Fleet Air Arm (FAA) of the Royal Navy had robust ideas about the application of helicopters – hunting and killing submarines – and had, in 1952, issued requirement NA.43 for an anti-submarine helicopter. The Sycamore HR12 was, in 1952, being evaluated for the role by the RN and RAF Joint Anti-submarine Warfare Development Unit at St Mawgan, with further work involving the Royal Aircraft Establishment (RAE) producing a Sycamore HR12 fitted with an AN/AQS-4 dipping sonar. Unfortunately, the Sycamore could not carry the sonar and a torpedo, so it was deemed too small for the FAA's NA.43 requirement, which called for a single hunter-killer platform.

As noted above, the Royal Navy acquired several Sikorsky S-55s, and when fitted with the AN/AQS-4 sonar, these served as Whirlwind HAS22s, but still as a hunter-killer pairing. Westland also produced the Whirlwind HAS7, powered by the Alvis Leonides Major, rated at 850hp (634kW) but these also operated in pairs, one Whirlwind fitted with a Rank-Pullin Type 194 dipping sonar, the other with a Mk.30 Dealer-B torpedo.

A Bristol Type 173, XH379, underwent trials on HMS *Eagle*, and the bulk of these involved deck handling. The FAA decided that such a machine was only suitable for aircraft carriers and, even then, would disrupt fixed-wing flying. (Blue Envoy Collection)

To fully meet NA.43, Bristol Helicopters proposed its Type 191, a twin-rotor, twin-engine type derived from the civil Type 173 – one of those civil transport projects mentioned in the Introduction that had taken up much the Ministry of Supply's helicopter development effort. Powered by two Alvis Leonides Major piston engines, the Type 173 was essentially two Type 171 Sycamore engine/rotor systems at each end of a new fuselage with a cabin for ten passengers.

The Type 191 was a shortened Type 173 with four-bladed rotors fitted, replacing the Type 173's three-blade rotors. To carry the required single 21in (53cm) Pentane or two Dealer-B torpedoes, a large ventral bay was installed and to allow access to this, the front undercarriage was lengthened. The AN/AQS-4 dipping sonar was to be fitted in the cabin along with operators' stations. The Admiralty ordered 65 Type 191s, plus three development aircraft, and to evaluate the type's suitability at sea, the prototype Type 173 (G-ALBN/XF785) was flown out to HMS *Eagle* to assess its deck handling. The trials would also provide experience of operating a large twin-rotor machine on an aircraft carrier.

The Type 191 suffered weight gain during development and, despite plans to replace the Leonides reciprocating engines with the lighter Napier Gazelle turboshafts, the weight still increased. Interestingly, the Type 191 was also on order for RAF Coastal Command as an inshore ASW platform to meet OR.346/Specification HR.149. Meanwhile, at the Admiralty, the view was that such a large helicopter could only operate from an aircraft carrier and would interfere with fixed-wing operations. This, compounded by the weight gain, saw the Type 191 cancelled in May 1955. The work Bristol carried out on the Type 191 was not wasted but carried over to the machine to meet Air Staff Requirement OR.325 – the Type 192 Belvedere HC1 medium support machine for the RAF.

Bristol's Type 173 matched two Leonides engines and main rotors from the Type 171 Sycamore with a new fuselage. The machine in the background, XN379, was used for naval trials on HMS Eagle. The Type 191 anti-submarine helicopter was to be shortened, powered by Gazelle turboshafts, fitted with four-bladed rotors and an extended front undercarriage. (Blue Envoy Collection)

Hunter-Killers – Submarines

The Admiralty had discovered during the Great War that the mere presence of an aircraft, any aircraft, could severely hamper the operations of U-boats. When RN officers first saw Sikorsky's helicopter, they knew exactly how to apply it – anti-submarine warfare from ships, especially small ships. The Admiralty aimed to replace the land-based, fixed-wing aircraft, of RAF Coastal Command or flying from escort carriers, that it had relied on during the Battle of the Atlantic, with its own helicopters on smaller ships that could accompany the fleet and be in the right place at the right time.

Unfortunately, early helicopters could not carry a useful payload and, as noted above, the second-generation machines such as the Whirlwind could not meet the Admiralty's NA.43 requirement. This called for a 'Single Package' hunter-killer machine equipped with the means to detect and destroy enemy submarines. Even a large twin-engined twin-rotor type such as the Bristol Type 191 or the initial Wessex HAS1 with Gazelle turboshaft struggled to meet NA.43, but as radar and sonar technology became lighter and more compact, the 'Single Package' became possible.

Naval Lightweights

As ever with weapons development, there were a number of dead ends. The Bristol 191 was one such dead end, but another involved the Fairey Ultra Light, a small, compact machine that the Army Air Corps had become interested in to provide reconnaissance, liaison and casevac duties to meet Specification HR.144, but it failed to prosper when the Saro Skeeter was adopted. The Admiralty examined the Ultra Light as a delivery vehicle for lightweight torpedoes to meet a new specification, HAS.191 issued in August 1958. This called for what was basically a long-range version of the anti-submarine mortars developed during World War Two but delivering a homing torpedo rather than depth charges.

The Ultra Light comprised a fuel tank that formed the fuselage onto which a two-seat cockpit, a Blackburn/Turbomeca Palouste gas turbine, tail assembly, landing skids and a two-blade rotor were attached. The Palouste's compressor provided high-pressure air to the rotor's tipjets where fuel was added to provide thrust and thus provided torque-free lift for the machine. The payload was a single Mk.43 lightweight homing torpedo, delivered under radio direction from sonar operators on the ship.

The compact Ultra Light could be accommodated on small ships such as frigates, and trials took place aboard HMS *Grenville* and HMS *Undaunted* during 1957–58 prior to the issue of HAS.191, but

Fairey's Ultra Light tweaked the curiosity of the Army Air Corps and Fleet Air Arm at different times. The Army wanted the Ultra Light as a light liaison and AOP machine, while the Admiralty was looking for a torpedo delivery vehicle. Ultra Light G-AOUK was one of six built before the project was cancelled. (Blue Envoy Collection)

Saro's P.531-0 XN334 was used for sea trials aboard HMS *Ashanti* in 1962. This machine was used to assess suction cups on the skids but these proved less than ideal, as XN334 slid off *Ashanti's* helideck. The now familiar castoring wheels were selected instead. (Blue Envoy Collection)

the requirement soon changed from the carriage of two, rather than one, Mk.43 torpedoes. Despite increasing the rotor diameter and changing the tip jets, Fairey's Ultra Light could not compete with a new conventional machine from Saunders Roe.

The Army had embarked on a project to increase the capability of its Saro Skeeter AOP12s by adding seats and increasing power by replacing piston power with a gas turbine. This led to the Saro P.531, which was developed into the Scout for the Army Air Corps. The Admiralty became interested, and Saro P-531-0 XN334 underwent trials aboard HMS *Ashanti* in 1962, with a further two P-531-0s acquired and extensively evaluated, including operations against a submarine, aboard HMS *Undaunted* during 1959.

These trials led to a proposal called MATCH (MAnned Torpedo-Carrying Helicopter) and the issue of Specification HAS.216 covering development of the P-531-0 into the Sea Scout HAS1, so named because the Army's P-531-2 derivative was to be called 'Scout'. On Westland's acquisition of Saro's helicopter interests, the Saro Sea Scout became the Westland Wasp and was, like the Scout, powered by a single Blackburn Nimbus turboshaft. The Wasp HAS1 entered service from mid-1963 and it performed very well in the anti-submarine role carrying a pair of Mk.44 lightweight torpedoes, again delivered under the direction of a ship's sonar.

The usual anti-submarine weapons load-out on a Wasp HAS1 was two Mk.44 torpedoes that would be delivered under the direction of the ship's sonar. Also of note is the angle of the wheels to ensure the Wasp could be manoeuvred on deck but not roll off the helideck. (Terry Panopalis Collection)

Other weapons carried by the Wasp included the WE.177A nuclear depth bomb, four Nord AS.11 or two AS.12 air-to-surface missiles (ASM). The AS.12 and the Wasp are credited with damaging the Argentine Navy submarine ARA *Santa Fe* during the Falklands Conflict in 1982.

In the mid-1960s, the Fleet Air Arm's helicopters acquired a new anti-submarine weapon in the form of the WE.177 Nuclear Depth Bomb (NDB). This 600lb (272kg) weapon was of variable yield (0.5kT for shallow waters and 10kT for deeper waters). When used in anger, the Wasps doors were removed (as seen here) to reduce the effect of blast as the Wasp sped away at top speed. (Terry Panopalis Collection)

The Six-Day War of 1967 had shown the Western powers the threat posed by a fast patrol boat armed with guided missiles. To counter these, Wasps were armed with the Nord AS.12 air-to-surface missile pending the arrival of a more capable weapon being developed as the CL.834. Wasp HAS1 XT780 of 705 NAS is carrying two AS.12s. (Terry Panopalis Collection)

The Wasp served until 1988 when the last of the *Rothesay*-class frigates was paid off. The Wasp also served with the navies of New Zealand, Malaysia, South Africa, Brazil and Indonesia. It pioneered several techniques for operating helicopters from small vessels, including castoring undercarriage and negative pitch rotors that push the aircraft onto the helideck. In addition to the two pilots, the Wasp could carry a winchman and a stretcher case for search and rescue (SAR)/casevac or three passengers. Vertical replenishment (vertrep) with underslung loads could be carried out, albeit on a limited basis.

The Wasp's replacement also had its association with an Army machine, but both machines came as part of the Anglo-French Helicopter Agreement of 1967 (see Appendix 3) whereby Westland Helicopters Ltd developed a battlefield helicopter and a shipborne anti-submarine helicopter for British and French service. This design was the WG.13 and would lead to a very successful machine in the Lynx.

During its design process, the WG.13 came in a variety of guises, but the WG.13V was developed into the shipborne machine that was to be capable of finding and classifying submarines, attacking these with lightweight torpedoes, surface reconnaissance and anti-ship missions against fast patrol boats (FPB). The WG.13, to meet joint requirement NGASR.3335 and Specification 273, entered Royal Navy

A quartet of Wasps overflies HMS *Dido*. The Wasp was a perfect 'match' for the *Leander*-class frigates. Under the MATCH (MAnned Torpedo Carrying Helicopter) system, the frigate's sonar was used to pinpoint the target and the Wasp delivered a homing torpedo to that location. (Blue Envoy Collection)

Above left: During April 1974, Westland's trials Lynx HAS2 XX510 conducted operations in the Bay of Biscay aboard Marine Nationale frigate *Tourville*. The series of trials evaluated the Lynx's compatibility with French naval vessels. (Westland via Terry Panopalis)

Above right: The BAC CL.834 Sea Skua was selected as the ASM for the naval Lynx and entered service in 1982, seeing action in the Falklands Conflict. Weighing in at 320lb (145kg), the Lynx could carry four Sea Skua or a mix of Sea Skua and torpedoes if required. Lynx HMA8 XZ236 has just released a Sea Skua, with the Redstart boost rocket motor firing. (Blue Envoy Collection)

service in 1981 as the Lynx HAS2. Owing to delays with the dipping sonar and other ASW equipment, the Lynx HAS2 was, like the Wasp, only able to prosecute submarine targets under direction of the ship's sonar. Once in service, continual upgrades saw the Lynx HAS2 systems upgraded, although the full ASW kit of sonobuoys and dipping sonar examined for the Lynx did not materialise.

The Ferranti SeaSpray radar was used to supply the 'surface picture' for the Lynx and was coupled with the British Aircraft Corporation (BAC) CL.834 Sea Skua ASM to provide the anti-FPB capability, with up to four Sea Skuas carried by the Lynx. The Lynx HAS2 made its combat debut in Operation *Corporate*, the Falklands Conflict, tasked with providing reconnaissance, ASW protection to the fleet and using the Sea Skua against Argentine surface vessels. The Lynx and Sea Skua were in action again during Operation *Granby*, the 1991 Gulf War, attacking Iraqi surface units and providing reconnaissance for the coalition forces. Similarly, the Lynx operated in Operation *Telic*, during the initial invasion of Iraq and the subsequent operations from 2003 until 2011.

Throughout the Lynx's long life afloat, 39 years, Westland produced a number of upgrades and modifications to keep the type current. The Lynx HAS3 introduced uprated Rolls-Royce Gem 42-1 engines and a gearbox to match the extra power. The last Lynx variant for the Royal Navy was the HMA8, derived from the Super Lynx 100. This introduced British Experimental Rotor Programme (BERP) rotor blades, the tail rotor from the Army's AH7, the SeaSpray antenna moved to a chin installation (for 360° coverage, but to save money the RN's installation on the HMA8 only covered 180°) and an electro-optical turret with GEC-Marconi Sea Owl Forward-Looking Infra-Red (FLIR) sensor mounted on a platform on the nose.

Above left: Never touch the winchman! At least not until they have been earthed. This winchman descending from a Westland Lynx HMA8 is about to be earthed by a special earthing tool with a cable attached to the ship's earth. On vessels lacking such tools, the winchman may be dipped into the water prior to entering the liferaft or boat. (Blue Envoy Collection)

Above right: The Type 23 *Duke*-class frigates were equipped with the Lynx HMA8 but could also accommodate the Merlin HMA2. Surface threats were detected and targeted with the SeaSpray radar in a chin radome and the Sea Owl infrared system in a turret on the nose. (Blue Envoy Collection)

Below: Westland Lynx HMA8 XZ719 holds station off the starboard side of USS *Mount Whitney*, flagship of the US Navy 6th Fleet. The sleek lines of the original HAS2 have been lost with the HMA8's relocated SeaSpray radome for 360° coverage and the turret for the Sea Owl thermal imager. (USN via Terry Panopalis)

Westland also produced numerous new and improved models such as the Navy Lynx 3 and Super Lynx 300, although the Ministry of Defence (MOD) showed little interest in most of these proposals. The Super Lynx 300 design led to Future Lynx, a common airframe for Army and Navy use that incorporated the latest manufacturing techniques, Rolls-Royce CTS800 engines, advanced avionics and weapons including pintle-mounted .50-calibre machine guns.

The naval aspects of Future Lynx were covered by the Surface Combatant Maritime Rotorcraft project and the outcome of this was the AW159 Wildcat HMA2 that entered Royal Navy service with 825 NAS in 2013. The Wildcat HMA2's main sensor is a new generation Selex SeaSpray radar matched with new weapons such as the MBDA Sea Venom and Martlet guided weapons in addition to torpedoes and depth charges. The Royal Navy has received 28 Wildcat HMA2s and the type has been ordered by South Korean and Philippine navies.

Left: Lynx HMA8 ZF582 lands on the Royal Navy's flagship, HMS *Ocean*, during a visit to Sunderland, the ship's adopted city. Both ship and helicopter were reaching the end of their careers with the Lynx replaced by the Wildcat and *Ocean* by *Queen Elizabeth*. (Author)

Below left: The Leonardo AW159 Wildcat HMA2 has replaced the Lynx HMA8. The Navy and Army Wildcat share the same airframe and engines but differ in equipment and sensors. (Peter Edwards)

Below right: Formed in 2001, the Black Cats (until 2004, the Lynx Pair) flew Lynx HMA8s until 2014 when the AW159 Wildcat was adopted. The pair's distinctive displays came to an end in 2019 when it became the Wildcat Demo Team and 'lost' a Wildcat. Solo displays will now be the norm. (MOD/Open Government Licence)

Naval Heavyweights

Having looked at the lighter end of the naval helicopter spectrum, the heavier end – shore and carrier-based machines – requires examination. The problems associated with the piston-engined Whirlwinds of the RAF and RN are described in Chapter 2, with the radical increase in capability offered by the change to gas turbine power being the ultimate solution.

The first turboshaft-powered production helicopter to enter service with the British armed forces was the Napier Gazelle-powered Westland Wessex HAS1 to meet Specification HAS.170. This entered operational service in July 1961 but was essentially a short-term solution pending development of improved systems; the HAS1 carried the Type 194 dipping sonar but was less than satisfactory in the anti-submarine role.

The Admiralty took advantage of the improvements made in ASW technology during the 1960s, and much of the HAS1 fleet was upgraded to HAS3 standard to meet HAS.227 by replacing the 1,450shp (1,081kW) Napier N.Ga.13 Gazelle with the 1,600shp (1,193kW) Gazelle N.Ga.18. Changes to the ASW equipment included fitting the Type 195 dipping sonar, new navigation and control systems and the Ekco EK10 radar with a distinctive dorsal radome that lead to the HAS3 being called the 'Camel'.

Right: The original Westland Wessex HAS1 could carry the dunking sonar and torpedoes. Wessex XS880 has deployed its sonar sonde and is carrying a Mk.44 torpedo on the port fuselage hardpoint. (Terry Panopalis Collection)

Below: The Wessex HAS3, such as XM838, embarked on HMS *London* (note the ship's crest aft of the cockpit) a *County*-class destroyer. The HAS3 was fitted with the Ekco EK10 radar and earned the nickname the 'Camel' because of the dorsal radome. (Terry Panopalis Collection)

In its primary role as an anti-submarine platform, the HAS3 could carry a pair of Mk.44 or Mk.46 torpedoes, two Mk.11 depth charges or a single WE.177A nuclear depth bomb. Serving from October 1967, the Wessex HAS3's place on the RN's ships was taken by the Westland Sea King from 1970, but it survived on the *County*-class destroyers, thanks to their peculiar side-entry hangar that the Sea King could not use! The Wessex HAS3 remained in service alongside the *County*-class with the last HAS3 unit being disbanded in early 1983.

Wessex HAS3s were operated around the world by five squadrons, but one example, XP142 aka Humphrey, is the most famous example. One of two HAS3s deployed to the South Atlantic, and operating from the *County*-class destroyer HMS *Antrim*, XP142 provided navigation assistance for two Wessex HU5 assault helicopters based on RFA *Tidespring* during Operation *Paraquet* to recover South Georgia from Argentine occupation on 21 April 1982. The two HU5s were to evacuate 15 special forces troops from the Fortuna Glacier. Both HU5s crashed in the terrible weather conditions, but Wessex XP142 and its crew made two trips to recover the troops and the HU5 crews.

Three days later, XP142 was involved in depth charging the Argentine submarine ARA *Santa Fe*, which would subsequently be attacked by HU5s and Wasps using Nord AS.12 wire-guided missiles. Thanks to its interesting three days in April 1982, Wessex XP142 currently resides in the Fleet Air Arm Museum at Yeovilton.

The reason for the Wessex HAS3's short career was the rapid advance being made in the capability of helicopters thanks to new machines designed for turboshaft power. One such design was the Sikorsky S-61 that came in two guises; the S-61R transport helicopter with rear ramp that was extensively used for SAR as the HH-3E Jolly Green Giant by the USAF and the S-61D with 'boat hull' and sponsons used for ASW by the US Navy as the SH-3D Sea King.

Wessex HAS3 XM836 of 737 NAS alongside the *County*-**class destroyer HMS** *Devonshire,* **which has another HAS3 on its helideck. The** *County*-**class was the last bastion of the HAS3. (Terry Panopalis Collection)**

This formation of 21 Westland Wessexes is but a fraction of the over 300 Wessex built for the Royal Navy, and the type served into the 21st century. (Blue Envoy Collection)

Westland had taken out a licence to build the Sikorsky S-61 along similar lines to this SH-3H of the US Navy's HS-12, based at Naval Air Facility Atsugi in Japan. As well as changing the engines to Rolls-Royce Gnomes and a new flight control system, the WS-61 Sea King was fitted with British ASW equipment. (Blue Envoy Collection)

It was the latter variant that Westland opted to build under licence from 1969 to meet NASR.358, issued in 1962 to cover a general-purpose medium helicopter. The naval aspects of NASR.358 (which came to dominate the requirement, prompting the departure of the Admiralty) called for an ASW helicopter and with the cabin volume and lift capacity provided by a pair of Gnome turboshafts, the WS-61 Sea King proved ideal.

Having withdrawn from NASR.358, the Admiralty, in June 1966, issued NSR.6429 to replace the Wessex HAS3 and eventually the HU5 in the Commando role. Westland embarked on the Sea King's development, and four Sikorsky SH-3Ds were acquired from Sikorsky in 1967 to facilitate development of systems. The biggest change was the replacement of the US General Electric T58 by its licence-built variant, the Rolls-Royce Gnome. The first Westland WS-61 Sea King HAS1 took to the air on 7 May 1969 and the type went on to serve for almost 50 years in a variety of roles. The last Sea Kings left RAF service in 2018, but two continued to fly under civil ownership to train German Navy Sea King crews.

Although the Sea King was the 'Single Package' ASW helicopter the Admiralty had sought for two decades or more, like the earlier Wessex, the machines would work best in pairs, with one maintaining contact with the target while the other attacked it. The Sea King could lift four Mk.44 torpedoes or two torpedoes and a WE.177A nuclear depth bomb in addition to the full suite of submarine detection and tracking systems.

In February 1970, 824 NAS became the first unit to operate the Sea King HAS1 and 56 examples were delivered before deliveries of new-build HAS2 (and converted HAS1s) commenced in 1976. The HAS2 introduced uprated Gnome engines and a revised transmission with a six-blade tail rotor. The subsequent HAS5 featured MEL Sea Searcher, later Super Searcher, radar with an enlarged dorsal radome, upgraded signal processing with support for sonobuoys whilst retaining the six-bladed tail rotor. These were further enhanced to produce the Sea King HAS6. A few HAS5 variants were stripped of their ASW fittings (apart from the radar) and assigned the SAR role as the Sea King HU5.

The replacement for the Wessex HU5 support helicopter was the Sea King HC4 to meet NSR.6101, usually referred to as the 'Commando'. The HC4 dispensed with the sponsons and, with the ASW

Soviet warships always attracted attention and *Zhdanov*, a *Sverdlov*-class cruiser, is being photographed by a Sea King HAS5. The HAS5's Super Searcher radar in the enlarged radome is the primary sensor for the detection and classification of surface vessels such as *Zhdanov*. (Blue Envoy Collection)

The HAS1 and HAS2 Sea Kings presented a greater threat to Soviet submarines than their Whirlwind and Wessex predecessors. Sea King HAS1 XV711 is carrying four Mk.44 torpedoes on fuselage hard points. HMS *Leander* forms a backdrop, and *Leander's* Wasp HAS1 can be seen in the distance. (Blue Envoy Collection)

equipment stripped out, could carry up to 28 troops and lift much of the Royal Marines' equipment, including light guns and Bandvagn tracked vehicles.

Experience in the South Atlantic prompted the conversion of two HAS2 to HAS2 (AEW) standard to meet NSR.6119 with the fitting of the Thorn/EMI Searchwater radar (a modification of the radar used on the Nimrod MR2 aircraft) and consoles to provide airborne early warning cover for the fleet. Subsequent AEW Sea Kings were designated as AEW2A, while the 1997 Cerberus upgrade with improved Searchwater and systems produced the Sea King ASaC7. The last Sea Kings to leave Royal Navy service were the ASaC7s, which were finally retired in 2018 and replaced in 2021 by the AW101 Merlin with Crowsnest equipment.

In addition to its service with the Royal Navy as an ASW asset for almost five decades, the Sea King provided SAR services with the RN and RAF. The Navy's grey and red HAR5s and bright yellow HAR3s of the RAF became a familiar, and usually welcome, sight around the UK.

The Royal Marines has a long relationship with its Dutch and US comrades, and interoperability is key to its Arctic operations. Sea King HC5 Commando from HMS *Invincible* (visible in the distance) lands on the US replenishment vessel USNS *Pecos* in 2002. (USN via Terry Panopalis)

Above left: Sea King AEW2 XV664 of 849 NAS has its Searchwater radome deployed. It retains the dorsal radome, which was deleted from XV664 when it was converted to AEW7 standard and ultimately to ASaC7 with the Cerberus system. (Blue Envoy Collection)

Above right: Westland Sea King AEW2 XV671 and HMS *Illustrious*, with Sea Kings and Sea Harriers embarked, in the background. One of the lessons from the Falklands Conflict was the lack of airborne early warning cover, which in the eastern North Atlantic was to be provided by fixed-wing AEW aircraft of the RAF. (Blue Envoy Collection)

Not the Sea King variant normally associated with a Royal Navy warship, the winch operators of RAF Sea King HAR3 XZ591 are keeping an eye on the officer (whose arms should be by his side) being delivered to HMS *Bristol*. Interestingly, the well for the Limbo depth charge mortars (behind the rating with the static discharge wand) was converted to a temporary swimming pool during a refit! It was subsequently plated over to become a helideck. (Blue Envoy Collection)

An Unlikely Seabird

Work on a replacement for the Sea King dated back to 1969 when Staff Target NST.6433 was issued to replace the Wessex HU5 utility and Sea King ASW helicopters. Westland produced several design studies to meet this Staff Target, including the sleek WG.24 of 1970 and WG.27, a clean-sheet design drawn up in 1973. A common medium helicopter for the RN and RAF was suggested as NASR.365 by the MOD but dismissed by the Admiralty, with the preferred option being referred to as the Sea King Replacement (SKR) to meet the newly issued NSR.6646. Westland responded with the WG.31, a twin-engine type that also featured a rear ramp, and the MOD selected this type for development as the SKR. By 1979, this had become the three-engined WG.34, and various dynamics testbeds had been built and tested.

Meanwhile, the Italian Navy's Agusta-built ASH-3D and ASH-3H Sea King fleets were getting long in the tooth and in need of replacement, so after comparing their requirements, the countries' admiralties decided that their needs were similar enough to use the same airframe. As a result, Agusta and Westland entered into an agreement to develop an ASW helicopter based on the WG.34, and in November 1979, the UK and Italian governments agreed to develop a new helicopter, with the work being co-ordinated by a new joint company called European Helicopter Industries (EHI).

The resulting EH101 made its first flight in October 1987 and, after a somewhat protracted development, entered Royal Navy service in June 2000 as the Merlin HAS1. It was later redesignated as the Merlin HM1, an odd name since the Merlin, the smallest of falcons, is a bird of moorland and mountain. These Merlin HM1s were updated to HM2 standard with new avionics, electro-optical systems and updated ASW equipment with the last modified Merlin delivered in 2016. The Merlin HM2 can also carry the modular Crowsnest AEW system to replace the Sea King ASaC7 fleet, but that proved more difficult to develop than first thought and finally entered service in early 2021.

Exercises hone the skills of both helicopter and submarine crews. Hunting submarines is what the Merlin was designed for, but rather than the Royal Navy *Trafalgar*-class boat seen here, the Merlin's targets were Soviet and Russian Navy boats. (Blue Envoy Collection)

Above left: Another view of EH101 PP5, the first naval Merlin, this time accompanied by PP3 G-EHIL, the offshore support variant of the EH101 and Sea King HAS5 XZ570, the testbed for the Blue Kestrel radar. (Blue Envoy Collection)

Above right: EH101 PP5 shows how such a large helicopter can be accommodated on a Type 23 frigate during its trials. The rationale behind the tailplane position is clear once the tailboom is folded. The chequered area at bottom right is the 'Heli Grid' for the anchoring system that the Merlin and Lynx use. (Blue Envoy Collection)

Below: EH101 PP5 was also involved in the integration of the Merlin systems with the airframe and weapons such as the Stingray lightweight homing torpedo. Stingray is the main ASW weapon on the Merlin, Lynx and Wildcat. It can also be launched from ships and fixed-wing aircraft. (Blue Envoy Collection)

The Merlin was designed to operate from ships as small as the *Duke*-class frigates as well as the *Queen Elizabeth*-class carriers, Royal Fleet Auxiliary vessels and assault ships. The latter also operate the Merlin HC4 Commando variant, transferred from the RAF and upgraded for naval operations.

One little-known role for helicopters is Fleet Operational Sea Training (FOST) for which the Royal Navy operates the Eurocopter AS365N Dauphin. These fly in support of the Navy organisation responsible for training crews to operate newly commissioned ships. Based at Newquay Airport, the two FOST Dauphins enabled instructors and other personnel to join ships at sea without lengthy transits by boat.

Right: The AW101 Merlin also serves as a replacement for the Sea King ASaC7 AEW system. Called 'Crowsnest', the system uses the Thales Searchwater radar and Cerberus mission systems as a kit that can be fitted to the Merlin HMA2. Crowsnest-equipped Merlins are to be embarked on HMS *Queen Elizabeth* and HMS *Prince of Wales*. (MOD/Open Government Licence)

Below: AW101 PP5's trials included operations from Type 23 frigates such as HMS *Northumberland*. Type 23s were designed with a large flight deck and hangar to accommodate a Merlin. (Blue Envoy Collection)

The British military operates at least six AS365N Dauphins, with two in support of Flag Officer Sea Training (FOST). The FOST Dauphins are used to transport officers to ships whose crews are undergoing training. Dauphin ZJ164 is seen here, about to land on HMS *Monmouth*. The other four Dauphins are used to support special forces. (MOD/Open Government Licence)

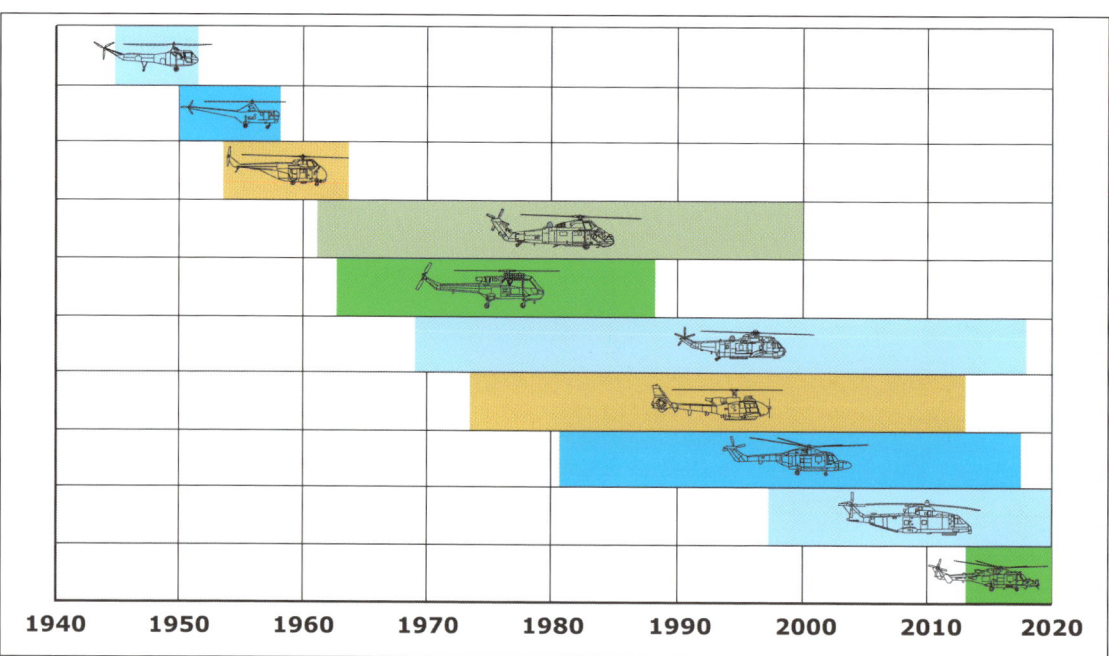

Timeline for the Royal Navy's helicopter. As with the RAF, the gas-turbine helicopter made the helicopter a practical and very useful machine. (Author)

Chapter 5
The Army Takes to the Air

Meanwhile, in the War Office, the British Army General Staff was keen to embrace the helicopter and had a list of roles that they considered the machines were more than suitable for. Aside from casevac and reconnaissance, the War Office wanted transports for which a larger, more powerful W.11 Air Horse seemed ideal as a replacement for the 3-ton truck. Unfortunately, the RAF had expressed the view that it should operate any flying machines intended for army support, be that ground attack or logistics, but did concede that the Army could operate machines lighter than 4,000lb (1,814kg).

The British Army, in 1948, began to look at fielding its own helicopters, evaluating the Sikorsky R-6 Hoverfly II in the liaison role, but suffered some bad luck in the acquisition of its first helicopters. Being available earlier than the Sycamore and having better performance than the WS-51 Dragonfly, Fairey's Gyrodyne was selected. After the prototype crashed in April 1949, the option changed to the Bristol Sycamore for use as a casevac and liaison type, but politics and economics intervened. After a dalliance with the Fairey Ultra Light and its tipjet-driven rotor, attention returned to a conventional design, and to facilitate a West German Army order, the Saro Skeeter was acquired to meet Specification H.163 for an army liaison machine within the 4,000lb maximum weight that had been agreed with the Air Staff.

Had it not been for the crash of the prototype, Fairey's FB-1 Gyrodyne would have been the British Army's liaison and utility type. Later converted to the prototype Jet Gyrodyne, G-AJJP survives in the Museum of Berkshire Aviation. (Via Terry Panopalis)

Two Saro Skeeter 6s were lent to the Army in early 1956 for evaluation against the Auster in the AOP role and found to be ideal, especially since the Skeeters could land close to the guns to brief the gunners. The British Army then acquired three Skeeter AOP10s and subsequently received its first helicopters in October 1956, with Saro Skeeter AOP12s replacing Auster AOP9s. This apparent delay when compared with the RAF and RN was mainly due to the Army being reliant on the other two services for its air support, be it ground-attack missions or airlift. The aircraft of the British Army's AOP squadrons were maintained by the RAF, but the pilots came from the Royal Artillery. Just after the war's end, HTF had trained 29 AOP pilots to fly helicopters, but since the Army had no helicopters at the time, they returned to flying Austers pending the arrival of the Skeeter.

The Skeeter AOP12 served with the Army Air Corps (AAC) from 1958 until 1967, while the RAF operated the Skeeter AOP12 and T13 trainer at Central Flying School to train the AAC pilots. The Skeeter provided the AAC with experience of operating helicopters and laid the foundations for the much more capable turbine-powered types that would follow.

The last of the piston-engined helicopters in the service of the British armed forces were acquired as gap-fillers pending the arrival of new turbine-powered designs, although the gap turned out to be rather longer than expected. To meet Specification AH.248, the Hughes 269 was proposed by Westland for the Unit Light Helicopter and Advanced Helicopter Training requirements, but the Bell 47G was preferred and selected for the Army (Sioux AH1) and the RAF (15 Sioux HT1). A total of 50 Agusta-Bell 47s were bought from Agusta in Italy and a further 250 examples were licence-built by Westland as the Westland-Agusta-Bell 47G-3B1. The Sioux AH1 was also used by the Royal Marines in 3 Commando Brigade Air Squadron (3 CBAS), which used machines, as it still does today, on permanent loan from the AAC. The first Sioux AH1s were delivered in 1965 and remained in service until 1978.

The Saro Skeeter 6 had been evaluated by the AAC and was particularly suited to the AOP role. The Skeeter gave the AAC valuable experience of helicopter operations. (Blue Envoy Collection)

Proposed for the Unit Light Helicopter and Advanced Helicopter Training requirements pending deliveries of the Westland Scout, the Hughes 269 would have been produced by Westland. This example, XS349, was evaluated prior to the selection of the Bell Sioux HT1 and HT2. (Via Phil Butler)

Intended as a stopgap, the Westland-built Bell 47 served with the AAC and RAF far longer than intended in the liaison and training roles. They were the last piston-engined helicopters in AAC and RAF service. (Blue Envoy Collection)

Chapter 6

Assault and Support by the RAF

Colonel Blacker's thoughts on 'Rotaplane Infantry' and the Director of Land/Air Warfare's paper on the Hover Force only came to fruition in the hills around the Ia Drang Valley in 1965. There had been earlier instances of the helicopter assault, including the first heliborne deployment of troops in Malaya on 16 February 1953 when a Dragonfly from 194 Sqn acted as a guide for three Sikorsky S-55s of 848 NAS, carrying four troops apiece, that set off to capture a terrorist leader. The first significant heliborne operation by British forces was conducted during the Suez Campaign, Operation *Musketeer*, of November 1956.

When used in support of the British Army, the Puma HC1 could carry troops and materiel such as vehicles and light guns where they were required. Three Puma HC1s of 230 Sqn approach a landing zone where vehicles await the troops. (Blue Envoy Collection)

One of the roles for the RAF's support helicopters on the Central Front was mining operations. Whirlwind HAR10 XP338 is delivering mines to a Bedford RL truck that is towing a Minelayer, Mechanical, Mk.1 with the pair being towed by a Caterpillar tractor. Whirlwind XP338 is fitted with a chute to lay mines on the surface. (Blue Envoy Collection)

On 6 November 1956, Whirlwind HAS22s (Sikorsky S-55s) of 845 NAS plus Whirlwind HAR2s and Sycamore HAR14s of the Joint Helicopter Unit transported 500 Royal Marines of 45 Commando from the carriers HMS *Theseus* and HMS *Ocean* to perform a heliborne assault on Port Said. The success of this (amid the political debacle that Suez became) led to helicopters being used for the deployment of troops and equipment, with the Whirlwinds replaced by the Bristol Belvedere HC1 in the RAF and modified Wessex ASW types in the Fleet Air Arm. The poor performance of the Belvedere saw it effectively banished to the Far East and replaced by the Wessex HC2 for the RAF, while the FAA acquired the Wessex HU5. The RAF subsequently acquired the Puma HC1, another outcome of the Anglo-French Helicopter Agreement, while the FAA took on the Westland Sea King HC4 Commando.

While not a true assault helicopter, the Boeing-Vertol CH-47 Chinook, via a long on-again/off-again saga, finally arrived in RAF service in 1981 and has been a key component of the RAF's support of Army and Royal Marines operations for over four decades. The Merlin HC3 was short-lived in RAF service but, on transfer to the FAA, replaced the Sea King HC4 Commando.

That first heliborne assault in Malaya laid the foundation for future helicopter support for British operations in Malaya and Borneo during the Indonesian Confrontation, from 1963 to 1966, and greatly influenced both helicopter operation and the types of helicopters used. One example of the latter is

One of the early model Bristol Belvederes, XG451, has the endplate tailplanes that were soon replaced by extended anhedral surfaces on later versions. Flying in the 1959 Farnborough Air Show, XG451 is lifting another of Bristol's wares, the Bloodhound Mk.1 surface-to-air missile. (Terry Panopalis Collection)

the conversion of the Whirlwind HAR4 from a disappointing piston-engined machine to the Gnome turboshaft-powered HAR10 that provided sterling service to the RAF for many years.

That the original 'lift' of four troops with the S-55 was deemed insufficient will be unsurprising, so a machine with greater capacity was specified, with 15 men or 5,000lb (2,268kg) of freight being considered ideal for the RAF. Being mainly aimed at jungle operations, the resulting helicopter would ideally be able to lift that payload straight up, out of a forest clearing under hot and high conditions. The Navy also required a similar capability to support the Royal Marines, and the events in Suez added to this desire for a more capable 'commando' machine to support amphibious operations.

Three Belvedere HC1s of 66 Sqn take to the air at RAF Seletar in Singapore. It could be argued that the Belvedere was a victim of its piston-engined antecedent, the Type 173, with the engines at each end of the fuselage. That installation precluded fitting a rear ramp, although design studies were in hand for such a configuration. (Terry Panopalis Collection)

Once in service, the Belvedere HC1 provided lift capacity lacking in the earlier machines as seen here with Belvedere XG456 lifting a L5 105mm pack howitzer. This view also shows one of the Belvedere's 'naval peculiarities' – the extended front undercarriage legs. These were to enable a torpedo to be loaded into the bay on the Type 191 from which the Belvedere was derived. (Blue Envoy Collection)

Meanwhile, the Air Staff, having dispensed with the need for a shore-based ASW helicopter, opted to use the Bristol Type 191 as the basis for a medium-lift troop-support helicopter. The resulting Bristol Type 192 Belvedere to meet the 1954 Specification H.150 was powered by two Napier Gazelle turboshafts, rated at 1,465shp (1,092kW). Unfortunately, the Belvedere came with 'RN peculiarities' and 'several uncorrected design faults' from the Type 191 including the Avpin engine start system that had a propensity to cause fires. The long undercarriage apparently suited the RAF who wanted 3ft (1m) ground clearance for landing in forest clearings. The other feature that hobbled the Belvedere was the location of the engines – within the fuselage and at each 'end'. The straight swap of the Leonides for the Gazelle did not utilise the full potential of the turboshaft in that they could be mounted where they had little impact on the cabin. The contemporary Vertol V-107/CH-46 Sea Knight used a far superior layout.

The Belvedere HC1 entered RAF service in 1961 and could carry 16 troops or 6,000lb (2,722kg) of freight, but it suffered from having a small cabin and a tall undercarriage that hindered the rapid embarking and disembarking of troops. The engine starting problems were solved by replacing the Avpin arrangement with a compressed air system developed from the Navy's similarly powered Wessex HAS3.

Despite this, the Belvedere served with distinction in the Middle and Far East (some might say it was banished) but was ultimately replaced by a much more capable and long-lived machine in the shape of the Wessex HC2 from 1964, with the last Belvedere retired in Singapore during 1969.

The Operation *Musketeer* assault involved RAF Whirlwind HAR2 support machines, and encouraged by this, the Fleet Air Arm stripped its Whirlwind HAS7 of ASW equipment and converted them to troop carriers for eight marines apiece. As the Wessex HAS1 replaced the Whirlwind ASW variants, the HAS7 'commando' variant of the Whirlwind entered service in 1960 and operated successfully in

Above: Belvedere's service exemplified in one photograph: XG468 on a hilltop amongst the barren rocks of Aden. The figures by the door give scale to the height of the cabin door, which did not ease the deployment of troops. (Blue Envoy Collection)

Left: Even without the exaggeration of the camera's perspective, the ground angle of the Belvedere HC1 and the high sill of the cabin door are evident in this photo of XG452. A few pilots broke ankles jumping from the cockpit when the starter system caught fire. (Blue Envoy Collection)

Borneo during the Indonesian Confrontation. The piston-engined Whirlwind HAS7 commando fleet was followed in 1962 by the Wessex Commando Mk.1, converted from Wessex HAS1s that had not undergone conversion to HAS3 standard.

Westland, in 1961, embarked on the development of a logistics/troop carrying/air-assault variant of the Wessex to address the troop-lift/air-support role for both services. The RAF specified a new helicopter based on the Wessex, but meeting Specification HU.224 required fitting the more powerful

Right: The Belvedere lacked the lifting capacity of its US contemporaries such as the Vertol CH-46 Sea Knight and, particularly, the CH-47 Chinook. Belvedere HC1 XG456 is lifting the fuselage of a Westland Wessex HAS1 Commando out of a jungle clearing. The Wessex has been stripped to bring it within the Belvedere's payload limits. (Blue Envoy Collection)

Below: With its air loadmaster striking a nonchalant pose in the doorway, Wessex HC2 XR525 lifts off from a landing site on the Otterburn Ranges, Northumberland. As an assault helicopter, the HC2 (and naval HU5) with the single door on the starboard side made disembarking 16 troops slower than on a machine such as the UH-1 Iroquois. The putative Wessex HC6 was to have doors on both sides. (Blue Envoy Collection)

de Havilland Coupled Gnome turboshaft, rated at 1,250shp (932kW) apiece but limited to 1,550shp (1,156kW) at the rotor head because of transmission limitations. This, however, gave the Wessex the twin engine safety required by the Air Staff, and the extra power enabled it to carry 16 fully armed troops or a 4,000lb (1,814kg) underslung load. The first two HC2s were converted from Wessex HAS1s and the first flight took place on 18 January 1962, with the first new-build HC2 making its maiden flight in October 1962.

Above: One benefit of the Wessex's maritime heritage was folding rotors and tail as shown by this RAF Wessex HC2 in West Germany. This made it compact and easily accommodated in a transport such as the Short Belfast C1. It could also be concealed in the field if required. (Blue Envoy Collection)

Left: The RAF's Wessex HC2s also operated in Norway in support of operations on NATO's Northern Front. Wessex XV721, in Arctic camouflage, lands to pick up stores including a pair of generators during an exercise in 1975. (Terry Panopalis Collection)

The type entered RAF service in February 1964, and the last of 74 examples was delivered in July 1965. The Wessex HC2 proved a reliable and rugged workhorse for the RAF in supporting the British Army wherever it operated, and the last Wessex HC2s in the RAF, used for SAR in Cyprus, were retired in 2003, while the last of the FAA's Wessex fleet, also in the SAR role, was retired in 1988.

Having operated the Wessex Commando Mk.1 since 1959, the Fleet Air Arm wanted a machine that was more powerful and capable than the Mk.1 and issued Specification HU.228. Ostensibly a Wessex HC2, the naval machine had a beefed-up airframe to withstand the rigours of low-level flight and a life at sea, it could also carry a wider range of weapons than the HC2. As an armed helicopter, the Wessex HU5 could be fitted with external weapons platforms on each side of the forward fuselage over the main undercarriage.

These armed Wessex HU5s could carry up to four AS.11 wire-guided anti-tank missiles, two AS.12 wire-guided anti-ship missiles, up to four 7-round 2in (5cm) rocket pods and Browning 7.62mm machine-gun pods. If the larger AS.12 was to be carried, a pair of 'mud guards' were required to protect the tyres of the main undercarriage from rocket efflux. The HU5 could also carry torpedoes and depth charges on the mid-fuselage hard points, if required, and external fuel tanks could be fitted on these hard points.

The Wessex HC2's army support role involved the rapid movement of troops around the battlefield. These troops (possibly Parachute Regiment) are formed up ready to embark on six Wessex HC2s of 18 Sqn. (Blue Envoy Collection)

The Fleet Air Arm's Wessex HU5s could be fitted with a variety of weapons, and WS501 has the full suite, comprising a 7.62mm machine gun, two 7-round 2in rocket pods (these are mounted back-to-back to form a cylindrical pod) and Nord SS.11 missiles. XS510 also carries a Mk.44 torpedo. This fit, apart from the torpedo, could be carried on both sides. The Wessex is fitted with wheel guards to protect the tyres from rocket efflux. (Blue Envoy Collection)

Left: The Wessex HU5 proved ideal for operations into the landing sites in the jungle-covered mountains of Borneo, acquiring the nickname 'Junglie'. Wessex HU5 XS479 of 848 NAS, HMS *Albion,* about to pick up a patrol. The rudimentary nature of these landing sites is evident, with tree stumps amongst the many hazards. Wessex XS479 is fitted with the external weapons platform and a GPMG in the cabin window. (Blue Envoy Collection)

Below: Wessex HU5s of 707 NAS landing on HMS *Hermes* in 1977. Seen from the front, the prominent exhausts either side of the nose are the main features for identifying the Gnome-powered Wessex variants. Three of *Hermes'* ASW Sea Kings can be seen ranged aft on the flight deck. (Terry Panopalis Collection)

The Big Cat from France

The Air Staff and Admiralty had been discussing a joint requirement for a medium helicopter to meet NASR.365 but, as noted above, the Admiralty abandoned a joint requirement to pursue the Sea King. The medium helicopter for what was now ASR.365 was to supplement, if not replace, the Wessex HC2 in the army support role, and amongst the types considered was the Bell UH-1 Iroquois. The problem with the Wessex in the air-assault role was its single cabin door on the starboard side, which made disembarking 16 troops less speedy than desired on a hot landing zone.

In the early 1960s, before the Americans became completely embroiled in the Vietnam War, aside from the British, the only country with experience of tactical air assaults using helicopters was France. The French Army's *Aviation Légère de l'Armée de Terre* (ALAT) had Sikorsky S-58 Choctaws and Piasecki H-21 Workhorses (essentially analogues for the Britain's Wessex and Belvedere) that it had used in Algeria and had embarked on replacing these American types.

The ALAT requirement called for fast deployment of troops from large cabin doors, plus increased speed and lift capacity by using all the available engine power, unlike the Wessex. The result was the Sud Aviation X-326A that followed the now standard 'box' layout. From this emerged the Sud Aviation SA 330 powered by two Turbomeca Turmo III turboshafts, rated at 1,185shp (884kw), all of which was available, unlike the Gnomes on the Wessex HC2. Couple that power with a sleek fuselage and

A pair of Puma HC1s, XW229 from 230 Sqn and XW201 of 240 OCU, resplendent in the original grey/green scheme applied to RAF helicopters in the 1970s and early 1980s. Such schemes would be replaced by drab green. (Blue Envoy Collection)

large doors each side (which halved the time to de-plane 16 troops compared with the Wessex), plus a retractable undercarriage, the result is a very capable assault helicopter. The SA 330A made its first flight in April 1965, which coincided with the first Anglo-French discussions on collaboration on helicopters (see Appendix 3).

Above: The Puma HC1 had undergone numerous upgrades and enhancements throughout its career, leading to a major upgrade that produced the Puma HC2 with Makila engines, glass cockpits and new systems including self-protection equipment. Puma HC2 XW212 shows many of the new sensors and self-protection systems on the airframe. (Peter Edwards)

Left: As a support helicopter for the Army, the Puma HC1 fleet found itself operating from dispersed sites. One drawback of the 'box' configuration of modern transport helicopters is access, as shown by the rigging gear needed to change an engine in the field. (Blue Envoy Collection)

The resulting Westland-built SA 330E Puma HC1 was more powerful thanks to its two Turmo III-C.4s rated at 1,300shp (969kW) and could lift 5,511lb (2,500kg) as an underslung load or up to 16 troops. The Puma HC1 entered service with the RAF in June 1971 and continued in service until the refurbishment of the fleet to HC2 standard was complete in 2015.

The RAF acquired 40 Puma HC1s and, rather than assault machines, these were used mainly for airlift of material such as light artillery but were particularly useful for supporting dispersed Harrier operations in the field. They were also a favourite mount of special forces, as the Pumas were fast and could alight in places other machines could not. The updated Puma HC2s remain in service, based at RAF Benson, and are earmarked for replacement under the New Medium Helicopter requirement. Several types have been mentioned, with the Leonardo AW149 being the front runner as of April 2021.

Another of the loads carried by Puma HC1s in the army support role was the 105mm Light Gun. This gave the troops and their support weapons much need mobility on the Central Front in what would have been a very fluid war of manoeuvre. (Blue Envoy Collection)

Puma HC1s could lift an external load of 5,511lb (2,500kg), which made them ideal for supporting dispersed operations in West Germany. The Land Rover seen here was an example of what could be carried, and lifting the equivalent underslung load was listed in the ASR.365 requirement. (Blue Envoy Collection)

Aérospatiale Westland Puma HC1s in the support role carry underslung loads during an exercise. Two of the Pumas are carrying long wheelbase Land Rovers fitted with the WOMBAT 120mm anti-tank gun, while the other two carry ammunition pallets. Also of note is the tail of Belfast XR368 and a 105mm pack howitzer. (Blue Envoy Collection)

One less than satisfactory outcome of the Anglo-French Helicopter Agreement was Sud Aviation's denial of authority for Westland to make further modifications to the Puma airframe, such as those required for carriage in transport aircraft. Unlike the Wessex, which took 24 hours to prepare for airlift, two Pumas could be readied in just over half the time, with four Pumas rather than two Wessex fitted into a Belfast. The original ASR.365 required the Puma to be prepared for transport in a C-130 Hercules C1 in a couple of hours, but in the absence of the requisite modifications, the entire upper structure and rotor head had to be removed in a lengthy procedure.

Right: The Puma is fast, and its manoeuvrability makes it well suited to operations involving places such as forest clearings that the RAF's other transport helicopter, the Chinook, cannot go. Puma HC1 XW223 of 230 Sqn makes a sprightly departure from an airfield. (Blue Envoy Collection)

Below: Westland Helicopters wanted to meet the RAF's requirement for transport of the Puma in the C-130 Hercules, but Aérospatiale denied Westland the necessary permissions to modify the Puma. As a result, only outsize cargo aircraft could carry them. Puma HC1 ZA934 is being unloaded from the hold of a USAF Lockheed C-5B Galaxy in Saudi Arabia during late 1990. (Blue Envoy Collection)

While the Puma HC1 successfully replaced the Belvedere HC1 and Whirlwind HC10, a replacement for the Wessex and Puma was sought in the mid-1980s. To meet AST.404, which did not come to fruition, a number of designs were examined including the Westland WG30, Westland-Sikorsky WS-70 (UH-60 Black Hawk re-engined with Rolls-Royce RTM322 engines) and Aérospatiale's AS322 Super Puma. In the end, the ultimate replacement for the Puma HC1 was the Puma HC2, with 24 HC1s re-engined with the Turbomeca Makila engine, rated at 2,414shp (1,800kW), upgraded dynamics and digital avionics. These refurbished HC2 machines could carry double the payload of the HC1 for three times the distance and, given the rise of expeditionary warfare in the early 21st century, be prepared for air transport in four hours.

One type that had a brief career with the RAF (but not as brief as the Belvedere) was the AgustaWestland AW101 Merlin HC3. Developed to replace the Fleet Air Arm's ASW Sea Kings, the Merlin was seen as neither fish nor flesh by an Air Staff that aimed to consolidate its fleet around Chinooks and Puma HC2s. The Air Staff's view was that the Merlin, in its tactical transport guise with a rear ramp, was too big for tactical work but too small for heavy lift and it was a reluctant customer for

Briefly considered as a Puma HC1 replacement, the AW101 Merlin HC3 did not serve long in the RAF, which instead opted to upgrade the Puma HC1, seen here in 2002, to HC2 standard. (MOD/Open Government Licence)

In the army support role, the AW101 Merlin HC3 was intended to fit between the Puma HC1 and Chinook HC2 and HC3. Seen in the company of an AAC Lynx AH9, this AW101 might be a pre-production example on trials, as it lacks the various 'lumps and bumps' of service HC3 machines. (Blue Envoy Collection)

the type. The RAF had a brief a dalliance with a combat search and rescue role for which three Merlins were modified. The modifications included an inflight refuelling probe and a winch, but the role was deemed unnecessary, and the kit was removed.

The first Merlin HC3s entered service in January 2001 and from 2003 they were deployed to Afghanistan and Iraq in the war on two fronts that soon found the RAF's helicopter capacity wanting. To increase the RAF's fleet as soon as possible, in 2007, six AW101 tactical transports in service with the Royal Danish Air Force were exchanged for six replacement AW101s on the production line. These former Danish machines were reconfigured for the RAF and entered service as the Merlin HC3A but remained in the UK as training machines to free up standard HC3s for operations.

The Merlin HC3 nearest the camera is very interesting, as it is one of three fitted with an in-flight refuelling probe, which is just visible forward of the nose. The intention was a combat SAR capability, but it was not pursued. (Blue Envoy Collection)

AgustaWestland Merlin HC4 Commando approaches a landing zone with its tail ramp and cabin door open. The Merlin HC4 and HC4A are conversions of former RAF HC3 and HC3As transferred from the RAF. The conversion involved fitting a folding tail and main rotor blades, cockpit from the ASW HMA2 and other 'naval peculiarities'. (Peter Edwards)

The Air Staff's preference for a Chinook/Puma fleet resurfaced in 2009 and from 2012 the entire RAF fleet of HC3 and HC3A Merlins was transferred to the Royal Navy to replace its Sea King HC4 Commando fleet. The fully converted Merlin HC4/HC4A Commando was fitted with folding rotor blades and other naval equipment, and the first fully 'navalised' Merlin entered service in May 2018.

Above: The RAF had acquired six AW101 Model 512 Merlins from the Royal Danish Air Force in 2007. These entered service as the Merlin HC3A and are identifiable by their nose radomes. Like the Merlin HC3s, these were subsequently transferred to the Fleet Air Arm for conversion to HC4A. This pre-conversion HC3A is about give a motorcyclist a fright! (Peter Edwards)

Left: The RAF's Puma HC2 fleet is due for replacement by 2025, and while a number of machines have been suggested, Leonardo's AW149 is being touted as the replacement. The infrastructure to support the AW149 is already in place, as the AW189 is the civil offshore support variant. AW189 G-OENC lands on the mobile offshore drilling unit *Ocean Patriot* in March 2021. (Author)

Object of Desire – The Snow Eater

The Air Staff had shown interest in a heavy-lift machine for Army support, and in the late 1950s, two types were available. Fairey's Rotodyne Z, a larger military version of the passenger transport being developed for British European Airways (BEA), offered excellent lift capacity, but the Air Staff was never in favour of the Rotodyne. Westland also offered the Westminster, a single-rotor machine that combined the dynamics of the Sikorsky S-56 (the US Army's CH-37 Mojave) with a new fuselage and Napier Eland turboshafts. On the amalgamation of Britain's rotorcraft companies, Westland opted to develop the Rotodyne, as it was government-funded and well into its development programme. The Rotodyne was subsequently cancelled because of arguments between the customers (BEA and RAF) over who should pay for development of the Rotodyne Z.

Right: Westland attempted to interest the British armed forces in a large transport helicopter and developed the Westminster. In addition to licences for the S-51, S-55 and S-58, Westland had a licence for the piston-powered S-56, the Sikorsky CH-37 Mojave, and used its dynamics coupled with Napier Eland turboshafts to produce the Westminster. (Blue Envoy Collection)

Below: Westland cancelled the Westminster in favour of the Fairey-designed Rotodyne that was being proposed to the RAF as a heavy support helicopter as the Rotodyne Z. With a capacious hold and a rear ramp, the military Rotodyne had potential, but when the civil Rotodyne was cancelled because of concerns about noise and costs, Rotodyne Z also foundered. (Blue Envoy Collection)

The change from piston to turbine power gave helicopters what was viewed as unlimited power, at least in theory. However, to transfer that power into lifting capacity required rotors and a transmission to drive them. The Wessex HC2/HU5 had shown that available power might not equate to usable power, but if one configuration showed the application of available power as usable power it is the twin-rotor helicopter epitomised by the Chinook. The name derives from the Native American Blackfoot tribe's name for the wind that melts the winter snows on the prairies east of the Rocky Mountains.

Unlike a conventional helicopter with main and tail (anti-torque) rotors, whereby a percentage of the power goes to driving the tail rotor, on the twin-rotor machine, all the power goes to the main rotors to lift the machine and its load.

Chinooks have a long and distinguished career in the RAF, with one machine elevated to war hero status, but the type's entry into RAF service had a long gestation, with the machine being an object of desire at the Air Staff since it first appeared on the scene in 1961. Disappointment with the Belvedere HC1 and its subsequent short career prompted the Air Staff and Admiralty to issue of NASR.358 in September 1962.

Four types were proposed, Westland's twin-rotor WG.1A that looked not unlike a Vertol V.107 Sea Knight but powered by four Gnomes, the Sud Aviation Super Frelon and two American types. The US machines would be the Sikorsky S-61R (CH-3C) and the Boeing Vertol CH-47 Chinook. The Super Frelon and S-61R were dismissed as being too big for the aircraft carriers and the WG.1A was a 'paper plane', which left the Chinook. Meanwhile, Sikorsky had substituted the S-65 (CH-53 Sea Stallion) for the S-61R, aiming to match the Chinook's lifting power.

To further tempt the British, Boeing Vertol drew up a Chinook powered by a quartet of Gnome turboshafts rather than two Lycoming T55s, but no British order was forthcoming. The Air Staff was, by 1964, backtracking on fielding such a large helicopter with a large payload that would essentially provide a juicy target for the enemy, nor could it fit in the Armstrong Whitworth AW681 transport aircraft. The policy of the time was to maintain British outposts East of Suez and reinforce these for 'interventions' in former colonies as required, so the preferred option was a smaller helicopter that could be airlifted to the east.

This prompted the inclusion of the requirement to ready the helicopter for airlift in NASR.365, issued in January 1965. NASR.365 was issued to cover a new medium helicopter for the Navy and RAF, which as noted above led to the separation of naval and air force requirements and the adoption of the Sea King and Puma respectively. This still left a gap for a machine with a greater lift capacity that could replace the Belvedere.

The Air Staff's perseverance on the Chinook paid off in March 1967 when 15 CH-47B Chinooks were ordered for the RAF. This was short-lived, as the Sterling devaluation crisis of the summer of 1967 caused the order to become prohibitively expensive and subsequently cancelled in November 1967. Just under four years later, with the Belvederes scrapped in Singapore, and the RAF, like all the British armed forces, now being NATO/Western Europe focused, 15 CH-47C Chinooks were ordered to meet a revised ASR.358 (Issue 2) – and promptly cancelled on cost grounds.

As part of that 1971 helicopter procurement process, the Aircraft and Armament Experimental Establishment (A&AEE) conducted trials of the CH-53D Sea Stallion and the CH-47C Chinook. Three examples of each were evaluated, including two US Navy standard CH-53Ds and a USAF HH-53C, plus two standard CH-47Cs and a 'UK Demonstrator' Chinook. The Sea Stallion lost out in these trials because it failed to meet one of the key requirements of ASR.358 – carriage of a 20,000lb (9,072kg) payload over a distance of 65nm (120km). The UK Demonstrator Chinook outperformed the Sea Stallion and the other standard Chinooks.

A third Chinook order was placed in 1978, but it would need to see off two competitors, including a new one that had appeared on the scene from Westland in the shape of the EH101. Determined to get its Chinooks, the Air Staff held discussions with the Admiralty and General Staff. These amounted to the Navy being told that if they did not interfere with the Chinook purchase, the Air Staff would toe the line on the new *Illustrious*-class carriers, while the Army was asked to specify an artillery piece that only the Chinook could lift.

The next phase involved the Minister of Defence and the Treasury, with the former being shown a diagram of the Chinook, EH101 and, for some reason, a Sea King, all *sans* rotors. The diagram showed that the Chinook was in fact smaller than the other two and would be easier to conceal in the field. The Treasury was fended off by the placing of an unfeasibly large non-refundable deposit on the Chinook order, so large that cancellation would be untenable. In October 1980, the first of 30 Chinook HC1s, essentially a CH-47C with elements from the latest CH-47D, was delivered to Boscombe Down. Mission accomplished – after almost two decades.

Right: This US Marine Corps Sikorsky CH-53E Super Stallion is lifting a US M198 155mm howitzer, which is similar to the FH70. When the Air Staff embarked on procuring the Chinook in 1978, the Army was asked what guns would need to be moved. The reply was that the FH70 155mm gun was the weapon, and the Chinook moved it as required, but in trials the CH-53D Sea Stallion failed to reach the required distance. (Blue Envoy Collection)

Below: 'Yes Minister, the Chinook is the smallest of the three, so much easier to hide on the front line.' After many attempts, the RAF finally acquired its Chinooks. A diagram comparing the sizes of the Chinook, Sea King and EH101 was shown to the minister, who agreed with the Air Staff on the Chinook. (Author)

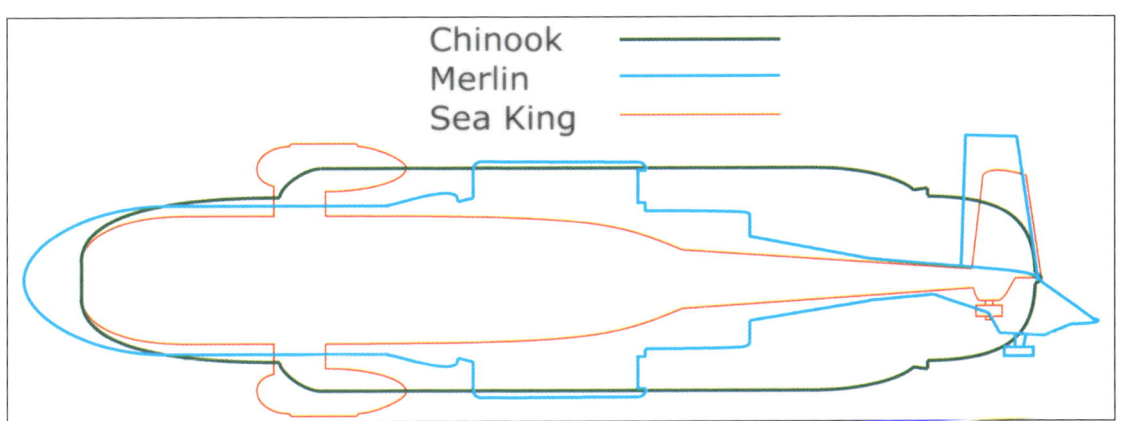

From its arrival in squadron service with 18 Sqn in August 1981, the Chinook fleet has been worked hard and within a year was at war in the Falklands. One Chinook, ZA718 (Bravo November), was the only Chinook to survive the Exocet strike on the SS *Atlantic Conveyor* in May 1982. Chinook HC1 ZA718 subsequently supported British forces in the recovery of the Falkland Islands, transporting troops and equipment, but has since operated in every theatre where the RAF's Chinook fleet served. Like much of the Chinook fleet, ZA718 was upgraded through the years and currently serves as an HC4. On a historic note, ZA718 has seen four of its crew win the Distinguished Flying Cross and the RAF Museum Hendon has an exhibit comprising the forward fuselage of a former US Army Chinook painted to depict ZA718, possibly a placeholder pending the actual aircraft's retirement.

Left: The first Chinooks in RAF service were HC1s, finished in the grey/green scheme used on most RAF aircraft in the 1970s and early 1980s. This Chinook HC1 ZA761 was delivered in November 1980 and has since gone through several upgrades and is currently an HC6A! The lifting capacity of the Chinook is demonstrated by ZA761 carrying a Spartan armoured personnel carrier. (Blue Envoy Collection)

Below: The Chinook HC1 was used for special forces support during Operation *Granby* in January–February 1991. This Chinook HC1 ZA677 was finished in a special scheme for night operations and, subsequently, did the rounds of the airshows during the summer of 1991. (Vic Flintham)

In a return to the original role of the RAF's first use of helicopters in Malaya, one of the critical roles performed by the Chinook is casualty evacuation, with the interior fitted out to evacuate critically wounded personnel to field hospitals within the 'golden hour'. The efforts of these crews cannot be praised highly enough.

The Chinook fleet has been upgraded, supplemented and converted for numerous roles including special forces support. The original HC1 fleet was upgraded to HC2 (the same standard as the US Army's CH-47D), while six were given a beefed-up forward fuselage to support an inflight refuelling probe to become HC2A. The HC3 was to support special forces along the same lines as the US Army MH-47E and included the enlarged sponsons. Rather than buy the MH-47E off the shelf, the MOD opted to produce their own cut-price special forces machine, the HC3. The resulting procurement debacle has gone down in history, mainly because the MOD did not include Boeing in the development of the digital flight control software. The resulting Chinook HC3, rather than being an all-weather machine, was unusable in any weather other than gin-clear conditions.

From 1998, the Chinook fleet was further upgraded, with the HC2s acquiring new avionics and engines to produce the HC4, while the blighted HC3s were similarly treated to become HC5s. With the Chinook's hot and high performance proving pivotal in the war in Afghanistan, a further 24 Chinooks were ordered, but the order was subsequently reduced to 14 examples as that war drew down. These were delivered as the HC6, comparable with the US Army's latest CH-47F, complete with digital flight control systems. This time designed and developed by Boeing.

Right: A Sikorsky R-6 Hoverfly II fitted for casualty evacuation on approach to land beside an army ambulance. From such humble beginnings grew the 21st century casevac Chinooks. (Blue Envoy Collection)

Below: In its 40 years of service, the RAF has operated a variety of Chinooks, including the HC6, which is equivalent to the US Army CH-47F. Chinooks, such as HC6 ZK552, have proved indispensable and have operated wherever British forces have been in action since 1982. (Peter Edwards)

Britain's Military Helicopters

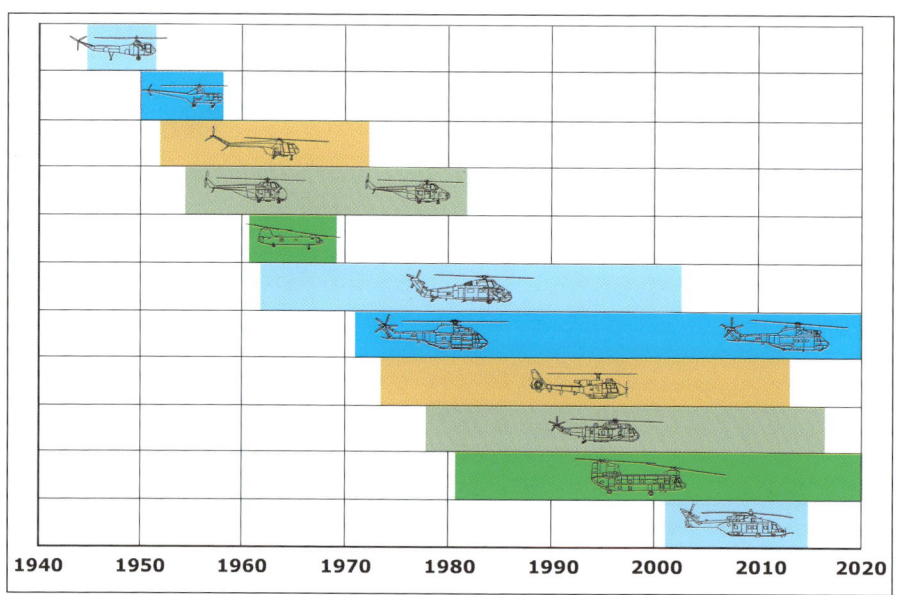

Since 2014, two types have formed the backbone of the RAF support helicopter fleet. The Chinook is a firm favourite with the RAF and the fleet has been upgraded on numerous occasions during the type's 40-year career. The Puma HC1 was upgraded to HC2 standard from 2012, making them modern and highly capable machines. (MOD/Open Government Licence)

Timeline for the RAF's helicopters. The longevity of the Puma against that of the Belvedere and Merlin is evident. Also evident is how the gas turbine-powered machines that entered service in the 1970s have stood the test of time. (Author)

Chapter 7
The Army and Its Helicopters

The 'feared enemy' is the Army Air Corps Apache AH1. Equipped with the Longbow system, it makes for a formidable fire support system with a battlefield management capability unsurpassed by any other attack helicopter in the world. The AAC's AH1s are to be remanufactured as AH-64E Apache Guardians. The blue colour of the Hellfire rounds on these Apache AH1s denote that they are drill rounds. (MOD/Open Government Licence)

Although latecomers to the helicopter, the British Army embraced them and now operates one of the most formidable battlefield machines in history in the form of the Apache AH1. Like the Royal Navy, the Army had sought a machine capable of fulfilling a specific role and similarly, this took many decades.

The Scouts
Having fielded the Skeeter, a more capable machine was in the AAC's sights. Initially a widened, turbine-powered Skeeter was considered, but, eventually, the Saro P.531-0 appeared (see Chapter 4) and this formed the basis of what became the Westland Scout and Wasp. Delays with the Scout's engine

In 1958, two Sud Aviation Alouette IIs were acquired for evaluation. These two, including XR232 shown here, were the first gas-turbine helicopters to enter service with Britain's armed forces and led to a further 15 being purchased. The type remained in service for almost 30 years. (Via Phil Butler)

prompted the Army to obtain a stopgap in the shape of a number of Bell 47s as the Sioux AH1. In addition to the Sioux, the AAC also acquired the Sud Aviation Alouette II, which became the first gas-turbine helicopter in AAC service when a pair arrived in 1958 for evaluation. A further 15 Alouette IIs were bought in 1961, and rather than being a stopgap, when the Scout finally arrived on the scene, the Alouette IIs carried on in the liaison role and were finally replaced by the Gazelle AH1 in 1988.

Even before the Scout had entered service, a new helicopter for the Army Air Corps was being planned by Westland designers who were examining a larger machine using more powerful engines that could carry an infantry squad under the designation WG.3. This design was too large, verging on the RAF's support helicopter such as the Wessex, so work ceased on the WG.3 in 1963. The French ALAT was looking for a light helicopter to replace the Alouette II, and in the spirit of Anglo-French co-operation of the time (1963, before the Anglo-French Helicopter Agreement was signed), Westland and Sud Aviation mated the Alouette II dynamics with the new Astazou engine to produce the WS-22. At the same time, Westland's engineers were also working on their own design called the WG.8 that was powered by the Allison T63 turboshaft.

The General Staff in May 1964 issued GSR.3336 for a new scout/liaison machine and Westland drew up the WG.10, a larger six-seater version of the WG.8, which was immediately dismissed as too expensive. Westland's next pitch was the WG.12, a smaller four-seater WG.10 using Scout dynamics and a 450shp (336kW) Turbomeca Oredon turboshaft. With little interest in this from the General Staff, Westland then proposed a Scout Mk.2 with the Astazou engine.

Meanwhile, across the Channel, Sud Aviation was working on a *Hélicoptère Leger et Observation*, analogous to the US Army's Light Observation Helicopter (LOH) requirement, for the ALAT. This appeared in design form as the X-300, which after modification was developed as the SA 340, better known as the Gazelle, that took to the air in April 1967.

With the signing of the Anglo-French Helicopter Agreement in January 1967 (See Appendix 3), the Gazelle became one of the three types to be jointly developed by Sud Aviation and Westland. The British Gazelles were to be built under licence by Westland but an initial order for 250 was

Left: The first production Scout AH1, XP846, is performing that perennial task of the utility helicopter – lifting stores. The Scout was a true utility helicopter, used for all sorts of tasks and the Army never ran out of new ideas for the Scout, such as Heli-Tele. One Scout was even fitted with a 7.62mm Minigun! (Blue Envoy Collection)

Below: Aérospatiale Gazelle AH1, ZB689, of 665 Sqn AAC cuts a dash on approach to landing. This Gazelle was one of a number fitted with the Heli-Tele surveillance system. (Peter Edwards)

subsequently reduced to 142. The first examples for the AAC were four SA 341 preproduction machines, while the SA 341B entered AAC (and Royal Marines) service as the Gazelle AH1 from July 1974. The SA 341C served with the FAA as the HT2 trainer, while 14 of the similar SA 341D were operated as the HT3 by the RAF. The RAF also planned to field the SA 341E as the Gazelle HCC4 communications helicopter but only acquired four examples.

Like the Scout, Gazelles served wherever the British Army and Royal Marines operated, and many of the AAC's Gazelle AH1s were fitted with the Ferranti AF532 Gazelle Observation Aid (GOA) to provide reconnaissance, observation and targeting information for Lynx AH1(TOW) machines. During the Falklands Conflict, some the Royal Marines' Gazelles carried SNEB rocket pods on their loading beams, but generally AAC Gazelles were unarmed. The Gazelle AH1 continues in service in the liaison and communications role but has been superseded in the training role by the Aérospatiale Squirrel HT1 and, latterly, the Airbus Juno HT1. The Gazelle might yet see 50 years in service.

Helarm Machines

As observed in Chapter 1, the Army pretty much knew exactly what it wanted of the helicopter as far back as 1940, before the helicopter, as it is recognisable today, had even flown. From these earliest ideas for an armed helicopter arose one of the most formidable weapons systems on the battlefield – the attack helicopter.

Given that the AAC had a 4,000lb (1,814kg) weight limit imposed on its helicopters by order of the Air Staff, there were few roles these light helicopters could fill, typically artillery observation posts, casevac and staff transport. The Skeeter AOP12 could not really fill the casevac role, but its replacement, the turbine-powered Scout AH1, could, and with an empty weight of 3,230lb (1,465kg) it was technically within the 4,000lb rule. However, with the capability to carry up to five passengers or 1,500lb (680kg) as an underslung load, it opened up a wider range of roles to the AAC. The Air Staff had grudgingly allowed the Army to fit 'bolt-on' weapons to its machines, obviously thinking that

The first flight of the prototype Lynx, XW835, took place at Yeovil on 21 March 1971. Thirteen prototypes and pre-production machines, both utility and naval versions, were built before the utility version entered AAC service in 1977. (Blue Envoy Collection)

the Skeeter, like its own Dragonflies, could carry nothing more lethal than a Sterling submachine gun poked out of a cabin window.

The AAC had other ideas, derived from a suspicion that when the Army called for armed air support, the RAF would be otherwise engaged. Given that the main threat to the British Army on the Central Front would be Warsaw Pact armour, an anti-tank capability would be useful, although the RAF saw that role as Air Force business with ground-attack aircraft. By the mid-1960s, anti-tank guided weapons (ATGW) were lighter and could be considered as bolt-on weapons for AAC helicopters.

Aside from pintle-mounted GPMGs in the doors, to convert the Scout for the helarm (Helicopter, Armed) role, the bolt-on weapons included some serious kit such as pods with seven 2in (5cm) rockets and 7.62mm machine guns as used on the Wessex HU5. Trials were conducted with the 7.62mm Minigun mounted in the cabin and firing out the port door, but the most formidable – and useful – weapon was the Nord SS.11 anti-tank missile.

The General Staff in 1968 had issued GSR.3431/1 for a helicopter-launched anti-tank missile, but the requirement was written around the Franco-German HOT (*Haut subsonique Optiquement Téléguidé Tiré d'un Tube*), as there was no British equivalent. Since the Scout could not carry the required eight rounds of the GSR.3431 weapon, an interim requirement, GSR.3492, was issued, leading to four SS-11s being carried by the Scout AH1. The bolt-on system comprised four SS.11 mounted on pylons attached to outriggers bolted onto the cabin, while an Avimo/Ferranti AF.120 sight was installed in the roof above the left-hand seat. The SS.11 installation gave the AAC an anti-tank capability on the NATO Central Front, while the full GSR.3431 system was being developed, but it was in another theatre that the Scout–SS.11 combination proved most useful.

For the first decade of its Army service, the Scout AH1 operated as a utility type in Borneo and Aden, where several Scouts were lost to enemy action. Into the 1970s, with 'withdrawal from Empire', the Scouts operated around West Germany and conducted operations in Northern Ireland, including surveillance with 'Heli-Tele' sensors. The machines 'on loan' to the Royal Marines 3 Commando Brigade Air Squadron (3 CBAS) were kept busy supporting commando operations in Norway and wherever the Marines were engaged, including the Borneo and Aden campaigns.

Then, in 1982, the Scouts became key elements in the recovery of the Falkland Islands and the bolt-on weapons proved invaluable. Six Scouts of 3 CBAS and six AAC Scouts were involved in Operation *Corporate*, mainly in resupply, casevac and insertion of special forces, but it was the SS.11 anti-tank missiles that proved pivotal to the operation in the last days of the campaign. A battery of Argentine

Lynx AH1 XZ172 and Scout AH1 XW280 in formation for a handover flight. Although the last Westland Scouts would soldier on in Brunei until 1994, the Lynx AH1 started to replace the Scout from 1979, and the SS.11-equipped Scouts began to be replaced by the Lynx AH1(TOW) from 1981. Apparently, Scout XW280 ended its days as a fish tank in the AAC base in Brunei. (Blue Envoy Collection)

Despite being bolted on, Nord SS.11 anti-tank missiles on the Westland Scout gave the British Army a potent anti-tank capability, independent of the RAF. Also visible in this view, above the port seat, is the Avimo/Ferranti AF.120 sight for the missiles. (Blue Envoy Collection)

105mm howitzers was holding up the advance on Mount Tumbledown, and the British infantry's mortars and Euromissile Milan ATGWs were outranged.

Three Scouts were fitted with outriggers and four SS.11 missiles apiece in a procedure that took 20 minutes each and was conducted with rotors turning. The howitzers' position was 'scouted' then engaged with ten missiles fired at a range of 3,000 yards (2,743m), resulting in nine hits on the guns, command post and their ammunition.

The helarm anti-tank Scouts were an interim solution, ideal for the 1960s and early 1970s, but as the Soviets began to field air-defence vehicles with their armoured formations, a longer-ranged, more capable anti-tank weapon was required to keep the helarm machine out of range of the air defences.

Hunter-Killer – Tanks

By the 1970s, with the focus now on the NATO Central Front in West Germany and the Soviet Horde with its massive tank armies, the Air Staff's diktat on the AAC's helicopters had faded into history. If the Warsaw Pact armies advanced across the Inner German border, rather than use tactical nuclear weapons as per the previous policy of massive retaliation, Flexible Response called for the use of conventional weapons, so every available anti-tank asset would be required. Unfortunately, as was shown in the October/Yom Kippur War of 1973, armoured formations now travelled with their own air defences, with the ZSU-23/4 *Shilka* and its four radar-directed 23mm cannon posing a particular problem for anti-tank helicopters such as the SS.11-armed Scout AH1.

The General Staff was determined to keep and update the AAC anti-tank capability and issued GSR.3431/1 to ensure the helarm machines remained outside the *Shilka's* engagement envelope of 3,280 yards (3,000m) 3,000m. The GSR.3431/1 range requirement was a minimum of 3,827 yards (3,500m), which greatly limited the choice of weapon and, more importantly, precluded the Hughes TOW (Tube-launched, Optically tracked, Wire-guided), which could not meet the 3,500m range requirement. The only British weapon available was the ground-launched BAC Swingfire, which was developed for helicopter launch by BAC. Known as Hawkswing, it proved to be too heavy and, thanks to its original launch regime, tended to hit the ground if launched at low level, i.e. when used against tanks. The General Staff issued a revised GSR.3431/2 that now included a newly developed extended range TOW (4,100 yards/3,750m) and combined with concerns about the effectiveness of the HOT warhead, the TOW system was adopted in August 1977.

The carrier for the TOW would be the British Army's new utility helicopter, the Westland WG.13, another of the types covered by the Anglo-French Helicopter Agreement. Through the early 1960s, Westland had been working on a replacement for the Westland Scout in the form of the Turbomeca Astazou-powered WG.3 battlefield helicopter. The WG.3a had grown from a machine to carry

A rooftop sight enabled the Lynx AH1(TOW) to acquire targets from behind cover. Lynx AH1(TOW) XZ607 has popped up to launch a FITOW missile but will drop behind cover while the missile flies to its target. Survival was a matter of launching and hitting the target before the *Shilkas* acquired the Lynx, so air defence elements were the first to be targeted. (Blue Envoy Collection)

ten troops or a 3,000lb (1,361kg) underslung load to the WG.3c that could carry 14 troops or an underslung Land Rover, which was approaching what the RAF viewed as its 'manor' owing to being in the Wessex weight class. This prompted a return to the lighter end of the scale with an all-up weight of 8,000lb (3,629kg) By 1967, the design had matured, but the Ministry of Aviation wanted a British or American engine, so the Continental T72 was selected, but ultimately the Rolls-Royce Gem, rated at 815shp (608kW), powered the new machine.

To separate this machine from the earlier WG.3 studies, Westland designated the new helicopter as the WG.13, which was proposed for the Army's requirement for a battlefield utility helicopter, GSR.3335, with Specification 268 issued in February 1969. After much discussion and reissue, the requirement became GSR.3335/3 with the accompanying Specification 269 for a utility version of the WG.13 issued to Westland in June 1972.

The first WG.13, designated Lynx AH1 in its army utility configuration, lifted off from Yeovil on 21 March 1971, and the first Lynx AH1 entered service with the AAC in August 1978. The utility Lynx, like the naval variant, underwent almost constant development throughout its four decades in service, with the Lynx AH5 (Interim) being fitted with a new gearbox and Gem-41 engines for evaluation by the RAE, although only two examples were built prior to the focus moving to the AH7, to meet GSR.3947.

The Lynx AH7 featured Gem-42 engines, a reinforced airframe, a new tail rotor (which rotated in the opposite direction to the original AH1) and various countermeasures to improve survivability. These

Lynx AH1(TOW) XZ179 (later converted to an AH7) is fitted with the mounts for the TOW missile launcher just aft of the cabin door. This Lynx is supporting anti-tank weapons trials as the load suspended under XZ179 is an Argocat all-terrain vehicle fitted with the Golfswing anti-tank missile system. (Blue Envoy Collection)

became the most numerous Lynx of the variants, and further improvements included night-vision aids and provision for night-vision goggles, developed using one of the Lynx AH5s.

Of particular interest was the adoption of the BERP rotor blade on the Lynx AH7. The British Experimental Rotor Programme (BERP) produced a new rotor tip shape that delayed compressibility on the advancing blade and stall on the retreating blade, thus enabling higher speeds. On 11 August 1986, Lynx Demonstrator G-LYNX achieved a world record speed for a Class E-1 helicopter of 216.5kt (400.96km/h).

The last Lynx variants in British Army service were the AH9, also known as the Light Battlefield Helicopter and the AH9A, the Light Assault Helicopter, intended for 24 Air Mobile Brigade. The AH9 dispensed with the original skids and adopted a fixed tricycle undercarriage that was more crashworthy and provided better protection for the crew and passengers. Powered by a pair of CTS800 engines, rated at 1,360shp (1,014kW), the AH9A had better hot and high performance and enabled the machine to carry greater loads, including the Browning M2 .50-calibre machine gun. The Lynx AH9A finally retired in January 2018, replaced by the Wildcat AH1.

Westland embarked on the British Experimental Rotor Programme (BERP) aimed at increasing helicopter performance, and the resulting BERP III blade did increase the forward speed of helicopters. A Lynx fitted with BERP III blades achieved a Fédération Aéronautique Internationale absolute speed record for a Class E-1 aircraft of 216.5kt (401km/h). (Blue Envoy Collection)

The Lynx AH9's tricycle undercarriage provided the Light Battlefield Helicopter with much-improved crashworthiness. This example, XZ170, also features the Hay Box infrared signature suppression system. The later Lynx AH9A featured CTS800 engines for improved hot and high performance. (Blue Envoy Collection)

TOW Lynx

For the helarm role, the Lynx AH1(TOW) could carry eight rounds, four on each side of the cabin, plus eight reloads in the cabin. The TOW missiles were sighted and tracked via a M65 stabilised sight mounted in the roof above the left-hand seat. The roof-mounted sight enabled the Lynx to acquire a target from behind cover such as trees or a ridge, lift slightly to launch the TOW, then drop behind the cover to track and guide the missile. The AH1(TOW) could also carry and deploy Milan anti-tank teams, with the two systems working successfully to counter an armoured assault. The other member of the Lynx helarm team was the Gazelle AH1, which, equipped with the Gazelle Observation Aid (GOA), could seek out targets for the Lynx.

Like the utility Lynx, the original 60 examples in the AH1(TOW) fleet were converted to AH7(TOW) standard. The Hughes TOW underwent regular updates, with Improved TOW (ITOW) featuring a stand-off probe to deal with spaced and reactive armour being the standard weapon when the Lynx AH1(TOW) entered service in 1981. Further Improved TOW (FITOW) carried a new warhead with a laser fuze, two self-forging warheads and a top-attack capability for use against the latest Soviet tanks such as the T-72. FITOW's warhead had been developed by Thorn EMI and the Royal Ordnance Factories and could be fitted to earlier versions of TOW.

When the AAC's anti-armour role transferred to the WAH-64 Apache AH1 from 2005, the TOW-equipped Lynx AH7 fleet aircraft lost their missiles but did retain the M65 sight and were redesignated as Light Utility Helicopters. The AH9 could not be equipped with TOW because of its wheeled undercarriage.

The Royal Marines' 3 CBAS also operated the Lynx in various roles, including anti-tank with TOW, from HM ships in support of operations wherever the Marines were deployed. The putative Lynx AH6 was a dedicated machine intended for the Royal Marines that was to use the same tricycle undercarriage, folding tail and rotors as the naval Lynx. Although better suited for shipborne operations, the AH6 was not built.

Left: Lurking in the trees, a Westland Lynx AH1(TOW) in its element. The helarm Lynx was to be used in ambush mode, over friendly territory and moving onto the enemy's line of advance as directed by Gazelle scouts. The Lynx could also deliver Milan teams that would act in concert with the Lynx force. (Blue Envoy Collection)

Below: The Lynx AH7 was an upgraded AH1 with Gem 41 engines and BERP main rotors. This AH7(TOW), XZ652, is fitted with TOW anti-tank missiles and the M65 roof-mounted sight but also carries the logo of the Blue Eagles, the AAC's display team. (Blue Envoy Collection)

The Lynx AH9A was well suited to hot and high operations in Afghanistan and this example, with a .50-calibre heavy machine gun in the port doorway, is providing the escort for an Afghan National Air Force Mil Mi-17 *Hip*. Also of note are the upwards pointing engine exhausts replacing the earlier 'Hay Box' type on the original AH9. (MOD/Open Government Licence)

The British Army and Royal Marines association with the Lynx lasted from 1979 until 2018 and, thanks to continual updates and technology such as BERP blades, the machine remained viable until its retirement. It still holds the speed record for a conventional helicopter.

Wildcat – a New Lynx

Over the four decades of Lynx development, Westland proposed some radical redesigns of the type, with design studies such as Lynx 3, powered by Gem 60 engines driving a rotor with BERP blades proposed for naval and army roles. The most visible change from the original Lynx was the rear fuselage that comprised a longer, deeper section tailboom, and the tail unit from the WG.30 transport replaced the Lynx tailboom. The cabin was also extended by 1ft (30cm) aft of the cockpit, and fuselage hard points similar to those on the naval Lynx were fitted, enabling carriage of a variety of weapons. The Lynx 3 elicited little interest but evolved into Super Lynx 300 (again with little interest from the MOD) and thence the Future Lynx.

The Future Lynx of 2002 caught the MOD's attention, as it would employ a common airframe and engines for Army and Navy roles, making it an ideal replacement for all Lynx variants then in service. The variant aimed at the British Army was focused on reconnaissance and thus became the Battlefield Reconnaissance Helicopter (BRH), with the MOD opting to order 40 for the AAC, although this order was reduced to 34 in 2009. The Future Lynx variant for the Royal Navy became known as the Surface Combatant Maritime Rotorcraft (SCMR) and an order for 30 examples of the SCMR variant was placed, later reduced to 28. The outcome of Future Lynx/BRH/SCMR was the AgustaWestland AW.159, known as Wildcat in British service.

As noted in Chapter 4, the Wildcat AH1 for the AAC and HMA2 for the Navy differ in role equipment, but it should be noted that the Wildcat benefits from the great deal of research that Westland conducted in the 1980s to produce an airframe with low observability – stealth – and ease of manufacture. This work was for a low-observable attack helicopter, the WG.47, but has resulted in the faceted fuselage and upward-pointing exhausts of the Wildcat.

The first Wildcat AH1 entered service with the AAC's Wildcat Fielding Squadron in May 2012, before entering full AAC service in August 2014. The Wildcat AH1 provides tactical support for the

British Army and Royal Marines and can be armed with a variety of weapons, including the Browning M3M, a helicopter-specific variant of the .50-calibre heavy machine gun.

Another utility type in service with the British Army is the Bell 212HP AH1, operated under a civil-owned/military-operated contract in Belize and Brunei. Powered by a Pratt & Whitney Canada PT6T Twin-Pac that comprises two PT6 turboshafts coupled to drive a common gearbox, which makes the Bell 212 ideal for hot and high operations in the jungles and mountains of Brunei in support of a battalion from the Brigade of Gurkhas.

Left: A rear view of an Army Air Corps Wildcat AH1 shows some of the survivability features developed by Westland during the 1980s. These include a faceted airframe and large upward directed engine exhausts pioneered on the Westland WG.44 and WG.45 studies for an attack helicopter. The tricycle undercarriage improves crashworthiness. (MOD/Open Government Licence)

Below: The AAC operates a number of Bell 212HP AH1s for operations in Brunei (in support of a Gurkha battalion) and Belize (British Army training). The power of the Bell 212 makes it ideal for supporting troops in jungle environments. The Brunei-based 212s replaced the last Westland Scouts in the AAC. (MOD/Open Government Licence)

Them... From Hereford

Oozing Italian style, the sleek Agusta A109 is more associated with music videos than a machine from which heavily armed special forces troopers are fast-roped. The Agusta A109A entered British Army service in 1984 after two Argentine Army examples were captured in the Falklands Conflict and put into service with 8 Flight AAC in support of 22 Special Air Service Regiment (22 SAS) operations. In the light of their suitability for SAS operations, these were supplemented by another two A109AM, and all four were finished in a variety of civilian-style schemes. The four A109s were replaced in 2009 with six Eurocopter AS365N3 Dauphins which, like the A109s before them, are based at Stirling Lines near Hereford. The Dauphins are also finished in civilian-style schemes and are now operated by 658 Sqn AAC in support of elements of 22 SAS.

Above left: Like the other three Agusta A109s of 8 Flight, ZE412 is devoid of any obvious identifying marking. The four machines were finished in a variety of civilian-style schemes, including this scheme on ZE412. This machine and another A109 were purchased to bring the unit establishment of 8 Flight to four aircraft. (Chris Lofting)

Above right: Agusta A109 ZE411 was one of two Agusta A109s captured in the Falklands Conflict and operated by 8 Flight, AAC. These were used in support of special forces and based at the Hereford base of 22 Special Air Service Regiment. (Chris Lofting)

The replacement for 8 Flight's A109s was the AS365 Dauphin, including ZJ782, to perform a similar role. In 2013, 8 Flight became 658 Squadron AAC, which also operates Gazelle AH1s in support of special forces. (Chris Lofting)

The Feared Enemy

One helicopter type missing from the British rotary wing arsenal until recently was the attack helicopter. Generally, what the media describe as a 'gunship' is an attack helicopter, best exemplified prior to 1991 by the Mil Mi-24 *Hind* but since Operations *Desert Storm* and *Desert Shield*, the media's favourite 'gunship' is the AH-64 Apache.

Even before the Lynx AH1(TOW) had entered service, the AAC realised that, because of the increasingly effective Warsaw Pact mobile anti-aircraft systems, the Lynx could not be operated over contested territory. The Lynx fleet was to operate on the friendly side of the forward edge of the battlefield area (FEBA), preferably like an ambush predator. Having seen the success of the US Army AH-1 Cobra in the anti-tank and escort roles in Vietnam, the General Staff, along with Westland Helicopters, began to examine the case for a similar machine in British service.

The initial response was a joint Anglo-German machine, the Westland/VFW-Fokker P277 of 1977, which used Lynx dynamics in a tandem cockpit airframe. Unfortunately, this machine was squarely aimed at a German Army requirement, and the General Staff had different ideas on the attack helicopter so turned its attention to the Agusta A129. Light, fast, agile and, best of all, cheap, the Agusta A129 Mangusta was the General Staff's choice over the heavier and more expensive Franco-German Tiger and the US Apache.

In July 1984, GST.3971 was issued for a Light Attack Helicopter and the requirement placed great emphasis on what the General Staff called 'survivability' with the main thrust of this being not to be seen, tracked or hit, and if hit, the crew should survive the inevitable crash. The Mangusta seemed ideal, aside from its restricted weapons load, four rather than eight Euromissile Trigat-LR missiles, but that could be fixed by more powerful engines.

Westland, meanwhile, had taken the low observable requirement to heart and produced a number of designs but had found the weight restrictions of GST.3971 difficult to meet. Eventually, in 1984, the WG.44 and WG.45 design studies were submitted for scrutiny, but like the other contenders (Eurocopter Tiger, Agusta A129 Mangusta and McDonnel Douglas AH-64 Apache) these had their shortcomings. The WG.44 and Mangusta were too small for the required missile load, while the WG.45

The Westland WAH-64 Apache AH1 was modified for operations from aircraft carriers and assault ships. These three Apaches are on the flight deck of HMS *Ark Royal*. (MOD/Open Government Licence)

was too big and therefore too expensive. Both the Westland machines introduced a faceted fuselage to reduce radar cross-section and exhaust systems that reduced the infrared signature. One subsequent study, WG.47b featured much reduced cockpit glazing to eliminate 'glint' and a twin tail rotor to improve manoeuvrability.

The AH-64 Apache was too heavy, lacked a mast-mounted sight and, because of its mode of operation, relied on a scout helicopter for targeting. Any Apache purchase would require scouts as well, with the Boeing–Sikorsky RAH-66 Comanche being the scout of choice. Unfortunately, in 1984, the Comanche was very much a 'paper plane'.

The next move by the companies was the creation of Joint Helicopter Industries Inc. (JHL) by Westland, Agusta, Fokker and CASA who proposed a radically modified Mangusta called Tonal. Despite being heavier and incorporating the Westland-developed stealth, the Tonal was looking ideal for GST.3971 as it stood in 1986, but the winds of change were blowing through. The Apache entered service in 1984 and was showing great promise and attracting interest from the British and Dutch armies. By 1990, great changes were afoot as the Soviet Union collapsed, the Warsaw Pact dissolved and Saddam Hussein's troops invaded Kuwait.

The TV War

By 1986, the General Staff was re-examining GST.3971. The Netherlands had departed the Tonal project, and soon the British did likewise and embraced the Apache, with British Army generals moving towards the American type. Then, in August 1990, Iraq invaded Kuwait and in the ensuing

The WAH-64D Apache AH1 can carry up to 16 Hellfires or CRV-7 rocket pods on its four pylons. The mast-mounted radome houses the antenna for the Longbow radar, used for target acquisition and surveillance. This example is at low level in the Mach Loop. (Peter Edwards)

Operation *Desert Storm/Desert Sword* of January and February 1991, the US Army's Apaches were in the vanguard of the advance, featuring in the extensive TV coverage. The world's armies were smitten by the Apache and it became the must-have attack helicopter.

Two years after the Apache stormed into the limelight, GST.3971 was superseded by Staff Requirement SR(A).428 that covered an attack helicopter for the AAC, and the McDonnell Douglas AH-64 Apache was on the list that included the Eurocopter Tiger, Bell/GEC Cobra Venom, Atlas Rooivalk and the Agusta Mangusta. Suffice to say the Apache won, in its AH-64D Longbow configuration that carried the AN/APG-78 Longbow targeting radar on a mast above the rotor head.

Westland built the Apache as the WAH-64D Apache Longbow with Rolls-Royce RTM322 engines and British avionics, plus modifications for use aboard HM ships. The first WAH-64D Apache AH1 entered service in 2005, and by May 2006, Apaches were supporting British forces in Afghanistan. Operations in Afghanistan prompted a number of upgrades to systems and increased ammunition capacity for the 30mm Chain Gun mounted under the forward fuselage. Known as the 'Monster' to the Taliban, the Apache certainly lives up to its name – Apache translates as 'the feared enemy' in the language of the Native American Zuni people.

The AAC received 67 Apache AH1s, and from 2019 these are being replaced with the most recent model Apache, the AH-64E Apache Guardian. These are remanufactured Apache AH1s, with the most obvious change being the replacement of the WAH-64D's Rolls-Royce/Turbomeca RTM322 turboshafts with General Electric T700. The weapons systems and avionics will also be upgraded, and the first Apache Guardians arrived in the UK in November 2020 and should be in full service by 2023. In the Apache the British Army has the close air-support system it had sought since 1942.

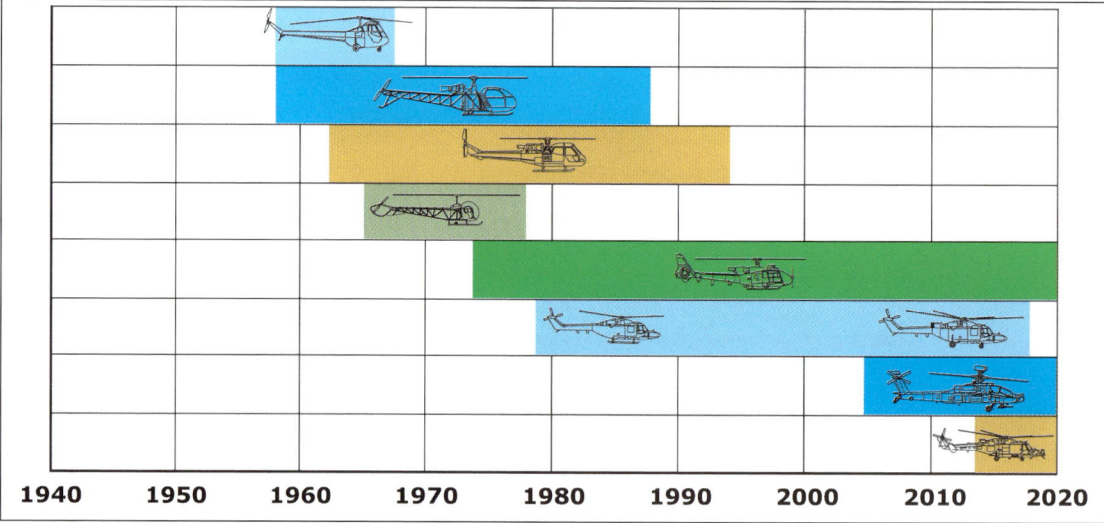

Timeline for the Army Air Corps' helicopters. The dominance of the Gazelle and Lynx shows how their continual development kept them in service. (Author)

Chapter 8
Rotary Training

The three services trained their helicopter pilots at many locations around the UK: RAF Shawbury for the RAF, naval pilots at RNAS Culdrose and Army Air Corps at AAC Middle Wallop. The RAF trained its first helicopter pilots on the Sikorsky R-4B and R-6, but the first designated helicopter trainers to enter service were the Skeeter T11 and T13, alongside AOP12s at Central Flying School Helicopter Wing (CFSHW) at South Cerney. Established in 1954, CFSHW used South Cerney as its base for helicopter training on Dragonflies, Sycamores and Whirlwinds as well as the Skeeters. In 1961, CFSHW moved to RAF Ternhill where it remained until moving to Shawbury in 1976.

These Aérospatiale AS350 Squirrel HT1s were based at Defence Helicopter Flying School, RAF Shawbury, where they provided basic training to pilots of all UK armed forces from 1997 until 2018. The Army Air Corps operated the Squirrel HT2 variant at its base at Middle Wallop. (MOD/Open Government Licence)

Saro Skeeter AOP6 XK773 (formerly Skeeter 6 G-ANMG) was one of the Skeeters evaluated by the British Army in 1955. XK773 is seen at RAF South Cerney, in Gloucester, home to the Central Flying School Helicopter Wing from 1961. (Via Phil Butler)

The RAF's No.2 Flying Training School (FTS) was established as a helicopter training establishment at RAF Ternhill in March 1976. Before the year was out, 2 FTS had moved to RAF Shawbury, and since 1998, all UK armed forces helicopter training was centralised at Shawbury at the Defence Helicopter Flying School, which has since become 1 FTS (No.1 Flying Training School).

As mentioned above, just after the war's end, the RAF had trained a few AOP pilots to fly helicopters, but without helicopters, the Army had little use for these newly qualified helicopter pilots who continued to fly Austers. Once the Army acquired its first Skeeter AOP12s, training commenced at CFSHW using Skeeter T13s, but by the 1960s, the Sioux AH1 was being used for training. These were replaced in turn by the Gazelle AH1s used for basic helicopter training before moving onto operational types such as the Scout and, later, Gazelle and the various Lynx variants.

The Fleet Air Arm had also used the Sikorsky R-4B and R-6 for training at RNAS Culdrose but from 1963, while the RAF and AAC acquired the Westland Sioux, the Fleet Air Arm operated the Hiller HT1 in the training role. A total of 21 examples, formerly US Navy Hiller HTE-2s, with 170hp (127kW) Franklin engines were acquired but only served for a short time. These were replaced from September 1962 by 20 of the more powerful Hiller HT2s, fitted with the 250hp (186kW) Lycoming engine. The last Hiller HT2 was finally retired in the spring of 1975 replaced by the Gazelle HT2 from 1974 and served until all helicopter training was centralised at Shawbury.

Left: The Fleet Air Arm's 705 NAS used Hiller HT1s like XB481 for basic helicopter training from 1953 until 1962. The Hillers replaced the Westland Dragonfly that had served as the FAA's basic training machine for helicopter pilots. (Blue Envoy Collection)

Below: The higher power Hiller HT2, such as XS165, replaced the Hiller HT1 from 1962 as the FAA's basic training helicopter. These were in turn replaced with the Gazelle HT1 from 1974. (Blue Envoy Collection)

The RAF's Aérospatiale Gazelles were the HT3 variant that flew in Central Flying School colours. The RAF's Gazelles were not replaced, with training being moved and centralised at the Defence Helicopter Flying School. (Blue Envoy Collection)

In 1997, the Defence Helicopter Flying School (DHFS) was established at RAF Shawbury to train pilots for all three services. Originally operated by FB Heliservices, the DHFS operation was taken over by Cobham and was equipped with the Aérospatiale AS 350BB Squirrel HT1 for basic training with the HT2 variant used for training AAC pilots at Middle Wallop. Conversion to heavier and more powerful multi-engined types was provided by the Bell 412EP Griffin HT1.

The MOD, in 2016, opted to privatise all flight training under the banner of the UK Military Flying Training System (UKMFTS), and Ascent Flight Training was selected as the prime contractor. Airbus Helicopters was to provide 29 Airbus H135s, designated as Juno HT1, and three Airbus H145s,

The Eurocopter H135 Juno HT1 replaced the AS350 Squirrel HT1 in the Basic Rotary Training role at the Defence Helicopter Flying School. All service pilots commence their rotary-wing training on the Juno, which, unlike the Squirrel, benefits from being twin-engined like all helicopters in UK service, apart from the three-engined AW101 Merlin. (MOD/Open Government Licence)

Left: As part of the UK Military Flying Training System (MFTS) programme, 202 Sqn operates the Eurocopter H145 Jupiter HT1 in the maritime and mountain training element of the Defence Helicopter Flying School. The Jupiters operate from RAF Valley and are equipped with a winch to train crew for types such as the Merlin HMA2 or Wildcat HMA2. (MOD/Open Government Licence)

Below: One of the roles performed by the AW139s of SARTU was training of crew in the various methods of using a rescue winch such as free winching or high-line. This latter technique is used in multi-survivor situations and involves a weighted line lowered into the liferaft or boat that will act as a guide for the subsequent winching operations. (MOD/Open Government Licence)

Above: Training in search and rescue was provided to military crews by the SAR Training Unit (SARTU) at RAF Valley. The unit flew the AgustaWestland AW139 (ZR327 in the photo) and Bell 412 Griffin HT1. FB Heliservices was contracted to operate and support both types. SARTU has been replaced by 202 Squadron flying Airbus H145 Jupiter HT1s. (MOD/Open Government Licence)

Right: Preparing trainee pilots for heavy, multi-engined helicopter flying, the Bell 412EP Griffin HT1 introduced pilots to the power they could expect with the Chinook, Puma or Sea King. The Griffins were operated by the Defence Helicopter Flying School until replaced by the Eurocopter H145 Jupiter in 2018. The Griffin HAR2 remains in service at Akrotiri in Cyprus. (Blue Envoy Collection)

designated Jupiter HT1. The latter, with 202 Sqn, are based at RAF Valley to provide SAR and maritime training for RAF and FAA pilots, although the SAR service has been privatised (see Chapter 10).

RAF Valley had been the base for the Search and Rescue Training Unit (SARTU) since 1979 and is tasked with training SAR helicopter crews in the specialist techniques that rescue operations over the sea and in the mountains require. The unit initially operated the Wessex HC2 and HAR2, but latterly the Bell 412 Griffin HT1 and AgustaWestland AW139 have been operated in the role. The unit was renamed as 202 Squadron and operates the Airbus H145 Jupiter HT1, while the Griffins and AW139s left Valley when the UKMFTS programme was introduced.

Learn to Test; Test to Learn

One establishment that leads the world in training is the Empire Test Pilots' School (ETPS) that was established in 1943 to develop formal techniques for testing new designs and to train pilots for testing duties. Since its inception, ETPS has operated a disparate fleet of aircraft, including helicopters, to provide students (always skilled pilots) with experience of a variety of types. The ETPS fleet has included Westland Lynx ZD550, Westland Sea King HC4X ZG829 and, more recently, Agusta A109E ZE416 and an Airbus H145. Future helicopter types destined for ETPS include the AgustaWestland AW139.

Left: The Empire Test Pilots School (ETPS) was established in 1943 to train the test pilots who conducted evaluation of new types. ETPS has operated several types of helicopter, including the Agusta A109 and, shown here, Westland Sea King HC4X, ZG829. (Blue Envoy Collection)

Below: The Fleet Air Arm's Aérospatiale Gazelle HT2 fleet was used for basic training. The instructors from 705 NAS formed a helicopter display team called The Sharks in 1975, which was popular at air shows around the UK. The team was disbanded in 1994. (Blue Envoy Collection)

Chapter 9
The Helicopter in Royal Service

As described in the Introduction, the Sikorsky R-4B was the first helicopter assigned to royal duties, with Hoverfly I KL110 (formerly of the Helicopter Training Flight) being attached to The King's Flight in 1947. For want of another role, the Hoverfly I delivered mail from Aberdeen to Balmoral. The first member of the Royal Family to fly in a helicopter was the Duke of Edinburgh in 1953, with the Duke becoming keen on helicopters and gaining his wings in a Westland WS-51 Dragonfly in 1956.

From 1947, a further four Sikorsky R-4Bs (KL106, KL973, KL987 and KL104) were at various times on loan from the Royal Navy, also for mail deliveries, but by 1954, a single Westland WS-51 Dragonfly HC4, XF261, was on loan for royal duties from Central Flying School. To replace the Dragonfly, the first of the more powerful and comfortable Whirlwinds commenced royal duties when Whirlwind HC2, XJ432, was lent to The Queen's Flight.

Specification HCC.127 was issued on 29 February 1958 and covered the bespoke Whirlwind HCC8 for The Queen's Flight and the first of two examples, XN126 and XN127, arrived on 5 November 1959. These differed from the other piston-engined Whirlwinds in RAF and RN service in being powered by the Alvis Leonides Major 160, rated at 740hp (552kW) rather than the Pratt & Whitney R-1340 of the HC2, Wright R-1300 of the HAR3 or Leonides Major 575 of the RAF's HAR5 and RN's HAS7. The HCC8s accommodated four passengers in its VVIP-configured cabin and were fitted with dual controls.

Sikorsky S-76C++ G-XXEB of the The Queen's Helicopter Flight, carrying the Duke of York, approaches for touch-down at the University of Hertfordshire. (Garry Lakin)

Whirlwind HCC8 XN126 served from 1959 until it was replaced by the Whirlwind HCC12 in 1964. (via Vic Flintham)

With the gas-turbine revolution, The Queen's Flight converted to gas-turbine power. In the spring of 1964, a pair of Westland Whirlwind HCC12s, XR486 and XR487, entered service with The Queen's Flight to replace the Whirlwind HCC8s. The HCC12 was a VVIP-configured variant of the RAF's Whirlwind HAR10 and HC10 that were powered by the de Havilland Gnome H-1000 turboshaft, rated at 1,050shp (783kW). Sadly, Whirlwind HCC12 XR487 was lost along with all four crew on 7 December 1967 because of a fatigue failure in the main rotor shaft, which prompted the grounding of all Whirlwinds pending main shaft inspection. A number of shafts were found to be defective and replaced, with the type returning to service, including royal duties, although according to official sources, Her Majesty The Queen did not fly in the type after the crash.

Ironically, XR487 was returning from a conference on its replacement when it crashed. The replacement was to meet Specification 267 and involved conversion of the Wessex HC2 to carry seven or ten VVIPs and included improved soundproofing, larger windows and a cabin step running the

Whirlwind HCC12 XR486 lands at RAF Wattisham for a royal inspection of 111 Sqn, whose Lightnings are lined up in the background. XR486 is preserved in The Helicopter Museum at Weston-super-Mare. (via Vic Flintham)

full width of the cabin door. Having ordered two VVIP Wessex aircraft in 1968, as the Wessex HCC4, XV732 and XV733 entered service with The Queen's Flight on 25 June 1969 with their first official duty in support of the Investiture of the Prince of Wales on 1 July 1969, although the Duke and Duchess of Kent had flown in XV732 three days before. Prior to the HCC4s entering service, a single Wessex HC2, XV726, performed training duties while on loan from 72 Sqn and as with previous types, HRH Prince Philip qualified as a Wessex pilot, as did HRH The Prince of Wales.

The Queen's Flight underwent major changes in 1995 when, on 1 April, it was merged with No.32 (The Royal) Squadron and moved from RAF Benson to RAF Odiham. April 1995 saw the end of dedicated royal duties by an RAF flying unit, although 32 Sqn still carries members of the Royal Family when required. The rotary winged element is called The Queen's Helicopter Flight (TQHF) and now operates from Northolt.

The Wessex HCC4s continued to serve in TQHF until 21 December 1998 when they were replaced by a single Sikorsky S-76+, G-AEXX, configured to carry six VVIP passengers. This S-76+ was retired in November 2009, with royal duties taken over by S-76++, G-XXEB, while a second S-76++, G-XXED was on the strength of 32 (The Royal) Sqn at Odiham. Another type used by the TQHF was Leonardo AW109SP GrandNew (probably better known as the Agusta A109SP), G-XXEC, on a long-term lease.

Another role for 32 Sqn helicopters was flying government personnel and senior staff of the services for which several types were used, including Whirlwind HC10s with VIP cabins. The Aérospatiale Gazelle was intended as a fast communications machine, but only four were assigned serials and one, converted from an HT3, entered service.

The Whirlwind HCC12 was replaced by the Wessex HCC4 in 1969 and served until 1998. Wessex HCC4 XV733 is resplendent in the gloss red and blue scheme that both machines were finished in. (via Vic Flintham)

Sikorsky S-76C++ G-XXED of TQHF touches down at Cheltenham Racecourse during the Cheltenham Festival in March 2020. (James Lloyds)

Following on from the Sikorsky S-76C++, TQHF was equipped with the Agusta A109E, including ZR323, seen flying over London. (MOD/Open Government Licence)

In addition to the HCC12 of The Queen's Flight, the RN and RAF operated several VIP helicopters, such as HC10 XP399 of 32 Sqn. These were fitted with five passenger seats (as used on the Bristol Britannia C1 transports) and ferried senior officers and staff. (MOD/Open Government Licence)

The communications variant of the Gazelle served with 32 Sqn at Northolt; Gazelle HCC4 XW855 was built in 1973 as an HT3 for the RAF. In 1978, XW855 was converted to an HCC4 and was in service as a VIP transport until 1996. (Via Phil Butler)

Chapter 10

For Those in Peril

The helicopters that are most visible to the general public are the machines involved in search and rescue around the British Isles. The RAF's bright yellow Whirlwinds, Wessexes and Sea Kings or the Royal Navy's red-panelled Whirlwinds, Wessexes and Sea Kings were rarely off television screens whenever the need arose, be it mass casualty events such as Piper Alpha or snowed-in farmers in need of fodder for their flocks. Since 2015, the Sea King, both the blue and the yellow, have been replaced by the red and white Sikorsky S-92s and AW189s of the Maritime and Coastguard Agency.

From the first encounters between British forces and the helicopter, one role recurred – rescue. That first encounter was in Burma on 23 April 1944 when the pilot of a RAF liaison aircraft and three crew members were airlifted to field hospitals by a US Army Air Force Sikorsky YR-4B. The formation of the Casualty Evacuation Flight (CEF) and its operations with Dragonflies in Malaya laid the foundations for the helicopter in the RAF's order of battle, with search and rescue being a key task.

From 1918, the RAF had operated a large fleet of boats as part of the RAF Marine Branch and its fleet included seaplane tenders, target-towing boats and rescue launches. It was the unlikely influence of an Aircraftman Second Class (AC2) Shaw that prompted the development of high-speed rescue launches in the late 1920s and early 1930s. Shaw had apparently witnessed a flying-boat crew drown because of the late arrival of a rescue launch, an event that prompted him to address the problem of aircrew recovery.

Visible on this view of Sea King HAR3 XZ594 of 22 Sqn are the key features of a SAR machine: rescue hoist above the cabin door and hemispherical visual search window. The SAR Sea Kings made regular appearances on TV news and could be the RAF aircraft the public was most familiar with. (Peter Edwards)

AC2 Shaw was none other than TE Lawrence, who had signed up to the RAF on at least two occasions under false names: Ross, then Shaw. Of course, people knew exactly who he was, but this was never mentioned. Interestingly, AC2 Shaw's commanding officer at RAF Bridlington claimed he only ever saw Shaw in full RAF uniform once – on the day he left the service in 1935. Shaw was very much left to his own devices at Bridlington where, working in concert with Hubert Scott-Paine of the British Power Boat Company, Shaw was instrumental in the development of the fast rescue launch. It could be argued that Lawrence/Shaw laid the foundations of the Air Sea Rescue Service in World War Two and the subsequent development of search and rescue (SAR) in the post-war era.

The fast rescue launches of the Air Sea Rescue Services (ASRS) held sway for over two decades, plucking ditched aircrew from the drink during World War Two. The ASRS was subsequently renamed as the RAF Search and Rescue Force and also operated aircraft such as Vickers Warwick maritime patrol aircraft equipped with airborne lifeboats and Lindholme Gear. To recover crew from their dinghies or airborne lifeboat, amphibians such as Supermarine Walrus and Sea Otters or flying boats were used if the sea state permitted landing; otherwise they waited for a launch or nearby ship.

Just after the war, in January 1946, the Air Sea Warfare Development Unit of Coastal Command acquired a Hoverfly I to be used for air–sea rescue trials at RAF Thorney Island. The Hoverfly was variously fitted with a Sproule Net to scoop up a survivor or a strop dangled on a line for the survivor to don and be lifted out of the water. Neither was very satisfactory; the Sproule Net could injure the person in the water, and the consensus was that a float-equipped Hoverfly was required to alight on the water to pick up the downed airman. The Royal Navy's FAA, with the propensity for aircraft ditching off carriers on take-off, favoured the Sproule Net, carried by Dragonflies on plane guard duty.

From the strop on a line, the logical development was a winch, but that would be a heavy piece of kit that would need an operator and, given the limited capability of the early helicopters, could make lifting a wet casualty a somewhat fraught procedure. By the 1950s, the helicopter was becoming a practical proposition in many fields, and despite the severe restrictions in all aspects of performance, SAR became a viable proposition. The arrival of the Bristol Type 171 Sycamore and the Westland Whirlwind would allow a crewman-operated winch to be fitted, enabling them to lift casualties. The Air Sea Warfare Development Unit received a Sycamore HR12, which was fitted with a winch and undertook trials in 1953.

The Bristol Sycamores operated by 275 Sqn became, in 1953, the first SAR helicopter unit in the UK and operated until they were replaced by the Whirlwind. The type was extensively used overseas, and here Sycamore HR14 XJ897 has set down on a Cypriot hilltop. (Blue Envoy Collection)

In February 1955, a dedicated SAR unit, 22 Sqn, had been established at Thorney Island, equipped with the Sycamore HR12, with these being supplemented and ultimately replaced by the Whirlwind HAR2 from June 1955. The squadron comprised a number of flights that were posted around the country at various RAF stations, and this would become the norm for RAF SAR units. A second squadron, 202, was a former meteorological reconnaissance squadron that became a SAR unit at RAF Leconfield in 1964. As with 22 Sqn, 202 comprised a series of flights located around the UK, with two aircraft based at each host airfield.

The Sycamore HR14 and Whirlwind HAR2 could be fitted with a winch, and the Whirlwind was large enough to accommodate the crew to operate it. In addition to a winch operator, trials soon revealed that retrieval of an unconscious or injured survivor was difficult without someone to put them into the strop. The solution was the winchman and a winch hoist with the capacity to lift two people at once. Another piece of kit that also added to the weight was a stretcher that had been intended for use in mountain rescues and modified for winching operations.

As with the other roles the helicopter took on, SAR benefited from the advent of the turboshaft, with the SAR Whirlwind HAR2 fleet being converted from piston to turbine power. The resulting Gnome-powered HAR10 was an improvement on the HAR2, and a further 68 HAR10s (and HC10 transports) were built. The first HC10s entered service with Transport Command in November 1961, followed by HAR10 SAR machines with 202 Sqn in 1964.

The arrival of oil and gas exploration in the 1960s saw an expansion of civil operations in the North Sea. The installations could provide refuelling facilities for the RAF's SAR helicopters. Westland Whirlwind HAR10 XD186 is on the deck of one of the platforms in BP's West Sole Gas Field, which was discovered in 1965. (Blue Envoy Collection)

Right: 'It's no fish ye're buying, its men's lives.' Sir Walter Scott, 1816. Fishing has always been a dangerous business and the SAR helicopter has plucked many a stricken fisherman from the sea or lifted the injured off a boat. The winchman from Whirlwind HAR10 XD188 is approaching a fishing vessel to assist the crew. (Blue Envoy Collection)

Below: The RAF's SAR squadrons were split into flights of four aircraft that were located at RAF bases around the country. These four Whirlwind HAR10s of 22 Sqn at Leuchars in Fife are flying over the Firth of Tay. The Whirlwinds arrived in 1955 and were replaced by the Wessex HAR2 from 1976. (Blue Envoy Collection)

The Royal Navy SAR units also received Gnome-powered Whirlwinds for SAR operations, designated as the HAR9, which entered service in the summer of 1966. The change to turbine power greatly enhanced the Whirlwind, which by the mid-1960s was fitted with new navigation aids and systems for homing into SARAH and SARBE location beacons.

The deployment of the Whirlwinds also coincided with an expansion of tasking for the RAF and RN SAR units. Their main role was the rescue of airmen and seamen from stricken aircraft and ships, but in the 1960s, the public took to the water and ventured into the hills for leisure activities that invariably got them into trouble. The RAF and FAA were usually called upon to help the civilian authorities.

Throughout World War Two, the RAF had suffered many accidents that involved aircraft flying into high ground and this 'unexpected foe' took its toll on aircrew. The RAF had, in 1943, established

The RAF's search and rescue squadrons' primary role was rescuing aircrew from aircraft that had come to grief. Helicopters such as the Whirlwind would ultimately replace the fast rescue craft used since before World War Two. The war had also heralded many advances in survival equipment including life rafts such as the Type MS.5, Mk.3. (Blue Envoy Collection)

the Mountain Rescue Service to rescue survivors and recover casualties from high ground in the UK and, later, around the world. The wartime RAF mountain rescue teams' operations continued into peacetime and became, thanks to specialist knowledge and intense training, highly skilled volunteers available for 'call-out' whenever required. The mountain rescue teams work closely with the RAF and RN helicopter units to aid the civil authorities in a range of tasks.

The SAR units of the RAF and FAA received a boost when the Wessex HAR2 entered service to replace the Whirlwind HAR10 in the mid-1970s, while the FAA fielded a SAR variant of the Wessex HU5 from 1964. RAF Wessex HC2s had provided temporary SAR cover in the Middle East in 1967 (where the bright yellow SAR colour scheme first appeared) and subsequently during the period that the Whirlwind fleet had been grounded (see Chapter 9), setting the scene for the type's use as the RAF's principal SAR helicopter.

The Wessex HAR2 was essentially a HC2 with SAR equipment, and despite the extra power and cabin space of the Wessex, the overall capability, especially in respect of weather limitations, was little better than the Whirlwind. The main benefit was the confidence boost of two engines on over-water operations in the Wessex HAR2.

The introduction of the Wessex HAR2 in 1976 coincided with a reorganisation of the UK's SAR provision, with RAF Finningley becoming the home of the Search and Rescue Wing. The Whirlwinds of 22 Sqn were transferred to 202 Sqn so that 22 Sqn could receive Wessex HAR2s, which the squadron operated in the SAR role until the mid-1990s.

With the Westland Sea King replacing the Wessex in the Fleet Air Arm, the Sea King was also adopted in the SAR role from 1971 with the modified Sea King HAS1s of 819 NAS at HMS *Gannet* (as HMS *Gannet* SAR Flight after 819 NAS departed in 2001) and 771 NAS at RNAS Culdrose (HMS *Seahawk*). These were replaced by the Sea King HAS5 from April 1988 as the FAA's Sea King fleet was upgraded. These machines were dedicated SAR machines and were devoid of ASW kit, enabling the carriage of more SAR equipment and survivors, with the bonus of increased fuel capacity for increased endurance over the ASW Sea Kings. Latterly, these HAS5 SAR machines became HU5s or HAR5s and were finished in a grey scheme with red panels and 'RESCUE' titles. As the busiest SAR unit in the UK, the *Gannet* SAR Flight provided cover across an area that extended from Fort William in the north to the Lake District in the south, Edinburgh and the Borders in the east, and to the west, Northern Ireland, and owing to international agreements, 200 miles west of Eire.

Replacing the Whirlwind from 1976, the Wessex HAR2 was essentially the HC2 support helicopter with SAR equipment. The twin-engined HC2s were preferred over the FAA's single-engined HAS3s that became available as the Sea King came into service. Wessex XV724 served with 22 Sqn until it was replaced by the Sea King HAR3. (Blue Envoy Collection)

To acquire a similar capability, the Air Staff issued Specification 290 for a dedicated SAR variant of the Sea King. These incorporated changes including moving the rear bulkhead backwards to provide an extended cabin, cargo hook for underslung loads and, of course, SAR equipment that included a 'boiling vessel' to supply hot drinks, which made it a popular transport while on other duties. After some revision to the Specification, the Sea King HAR3 entered RAF service in 1978, replacing the last Whirlwind HAR10s and the yellow Sea Kings became a familiar sight around the country. The military SAR units co-ordinated with local police for land-based SAR such as mountain rescues or with HM Coastguard for maritime emergencies. Operations were handled by two regional Aeronautical Rescue Coordination Centres, one at Plymouth and the other near Edinburgh.

Sixteen HAR3s were initially delivered to meet the RAF's SAR requirements, with a further three ordered to provide SAR coverage in the Falkland Islands, although these were painted in a grey scheme. An additional six Sea King HAR3A variants were ordered to replace the last of the Wessex HAR2s from 1996 and featured digital systems and modernised navigation systems.

The last bastion of the SAR Wessex was Cyprus, where former FAA Wessex HU5s were operated by 84 Sqn RAF from Akrotiri. These were replaced by the Bell 412 Griffin HAR2 and continue to provide SAR coverage for the eastern Mediterranean.

The Royal Navy's SAR fleet included Sea King HAR5, XV661, from 771 NAS based at RNAS Culdrose. The Navy's SAR machines had all ASW kit removed but retained the Sea Searcher radar to provide a surface picture of shipping. (Peter Edwards)

Seeking to replace the Whirlwind and Wessex SAR fleet with a more capable machine, the Sea King was selected. The resulting Sea King HAR3 was an all-weather machine that was a welcome sight to many a mariner or climber. (MOD/Open Government Licence)

Left: The Sovereign Base Areas at Akrotiri and Dhekelia on Cyprus are provided with SAR cover from RAF Akrotiri. The Westland Wessex provided the cover from June 1984 until it was replaced with the Bell 412 Griffin HAR2 in 2003. Wessex XS485, seen here with a BAe Harrier GR5, is a former FAA Wessex HU5C converted to the SAR role and operated by 84 Sqn. (Blue Envoy Collection)

Below: The RAF's SAR responsibilities in Cyprus are met by 84 Sqn and its Bell 412 Griffin HAR2s such as ZJ704. In addition to SAR, the Griffins have been called upon to work as firefighting machines. (MOD/Open Government Licence)

Private Practice

From 1986 until February 2016, the RAF Search and Rescue Force and the Royal Navy's SAR Flights provided the United Kingdom with a search and rescue capability that was second to none. In 2006, HM Government opted to privatise the UK's search and rescue service and launched the Search and Rescue Helicopter (SAR-H) programme, essentially to spare the expense of replacing the SAR Sea Kings. A consortium comprising Sikorsky, Thales, Royal Bank of Scotland and CHC Scotia was selected in February 2010 to operate the SAR service under the name Soteria SAR using 24 Sikorsky S-92s. In February 2011, the Soteria consortium revealed that it had been given access to 'commercially sensitive information' regarding the bidding process. This prompted the reopening of the bidding process and the contract was awarded to Bristow Helicopters on 26 March 2013.

Bristow would operate 22 helicopters on behalf of the Maritime and Coastguard Agency (MCA), 11 Sikorsky S-92s and 11 AgustaWestland AW189 at ten bases around the UK, with two machines at each base (with two undergoing maintenance at any one time). Bristow had experience of SAR operations, having operated Sikorsky S-61Ns for the MCA from 1983 until 2007, when CHC Scotia took over with AW139s and S-92s.

Bristow's SAR machines, resplendent in their red and white scheme with 'HM COASTGUARD' titles, began to take over from the RAF and RN in April 2015, with the handover from the military units taking place over six months until the last RAF Sea King unit, A Flight of 22 Sqn, was stood down on 5 October 2015. The Royal Navy's last Sea King SAR unit, 771 NAS was decommissioned on 26 March 2016.

Since 2016, Bristow Helicopters has provided SAR cover to the UK, replacing the RAF and RN Sea Kings. Finished in HM Coastguard colours, the Sikorsky S-92 (and the AgustaWestland AW189) are flown by civilian crews. (MOD/Open Government Licence)

Conclusion

Since 1945, the helicopter has gone from a novel contraption of limited capability to a key machine on any battlefield be it land, sea or, indeed, air. They do go where nothing else can, and once the gas turbine became the powerplant of choice – surely the key event in their development – helicopters went from strength to strength. From one Whirlwind airlifting four soldiers armed with Lee-Enfields and a Bren gun during the Malayan Emergency to the assaults by air-mobile formations complete with troop and cargo transports under escort by heavily armed attack helicopters, the helicopter has changed the way wars are fought. At sea, they have revolutionised anti-submarine warfare by making the transition from Hoverfly I as a visual deterrent and light stores transport to fully capable anti-submarine platforms in the Merlin HMA2. The one role where they are appreciated by civil and military alike is search and rescue. The bravery and commitment of the SAR crews is without doubt, and if any military helicopter was in the public eye, it was the colourful machines of the SAR units.

Helicopters truly are wonderful machines whose full potential has yet to be met. They have seen off various vertical take-off and landing systems, and there is little doubt that they are here to stay. What will they do next?

To the Ends of the Earth. A leopard seal watches Westland Lynx HAS3(ICE), which carries a stabilised camera in support of the BBC crew filming *Frozen Planet* in Antarctica. Britain's military helicopters operate from the Arctic to Antarctica, where the Royal Navy's ice patrol ship HMS *Endurance* supported British Antarctic Survey research stations on the Antarctic Peninsula. Helicopters really do go where nothing else can. (MOD/Open Government Licence)

Appendices

Appendix 1 – The Westland/Sikorsky/Aérospatiale/Boeing Agreements

At the war's end, Westland Aircraft opted to end development of fixed-wing aircraft and concentrate on the emerging technology that was the helicopter. The Chief Designer, WEW 'Teddy' Petter, departed for English Electric at Preston and went on to design the highly successful Canberra and the P.1 that formed the basis of the Lightning. Westland, while conducting design studies of its own from scratch, also took licences for Sikorsky designs and this proved very successful indeed. Licences were acquired for the S-51, S-55, S-56, S-58 and S-61 with the ultimate Sikorsky licence being for the S-70 Blackhawk. The Westland-built types invariably featured British engines and equipment, adding the 'W' prefix to the Sikorsky designation.

Westland Helicopters also acquired a licence for the Boeing (née Hughes) AH-64 Apache, built for the Army Air Corps as the WAH-64, the 'W' prefix added to the US Army designation for the Apache rather than the manufacturer's designation.

Aérospatiale (in its previous guise as Sud Aviation) and Westland had collaborated long before the Anglo-French Agreement of 1967. The WS-22 of 1962 was one project that led to the Gazelle. It is possibly the earliest instance of Anglo-French co-operation on helicopters but foundered because of disparate requirements from the respective armies.

Westland has had a series of owners/collaborators including shareholdings by Sikorsky and Fiat, but in 1994 became wholly owned by GKN, which in 2000 agreed to merge with Finmeccanica to form AgustaWestland. GKN sold its share in AgustaWestland to Finmeccanica, and the company became part of the Leonardo Group in 2016 and became known as Leonardo Helicopters.

Westland obtained a licence to build the S-70 (US Army UH-60 Blackhawk) with the intention of interesting the MOD in buying the WS-70 as a Wessex replacement. This WS-70, ZG468, acted as a demonstrator for weapons carrying trials. (Blue Envoy Collection)

Appendix 2 – The Consolidation of Helicopter Companies

By 1960, the United Kingdom had four companies designing and building helicopters: Bristol, Fairey, Saunders Roe (Saro) and Westland. Percival Aircraft had dabbled in the helicopter, but its P.74 quite literally never got off the ground.

With the drive to consolidate the British aviation industry, the fixed-wing interests were placed into two companies, British Aircraft Corporation and Hawker Siddeley Aviation, plus Handley Page who refused to comply with the merger policy. The four helicopter companies were merged to form Westland Helicopters Ltd in 1961.

Appendix 3 – The Anglo-French Helicopter Agreement

Signed in January 1967 by the Rt Hon. John Stonehouse, Minister of Aviation, in the spirit of European solidarity extant at the time but also, and more likely, because of the desire to share development costs, this agreement brought France and Britain together to develop helicopters. Stonehouse famously did a 'Reggie Perrin' and faked his own death in 1974 and was caught because the Australian police thought he was the fugitive Lord Lucan! After his actual death, Stonehouse turned out to have been in the pay of Czech intelligence since the early 1960s. Interestingly, Stonehouse had championed the acquisition of Super VC10s rather than Boeing 707s by BOAC!

The Anglo-French Helicopter Agreement covered the development of three types for service with British and French armed forces. The Agreement covered licence production by Westland of the SA 330 battlefield support helicopter for the RAF as the SA 330E Puma HC1 and the SA 341 as the Gazelle AH1, HT2, HT3 and HCC4, while France would build the Lynx under licence.

In the end, the UK acquired over 200 Westland-built Gazelles of various marks and 48 Puma HC1, also built by Westland. France took 40 naval Lynx. All built by Westland.

The French types involved in the Anglo-French Helicopter Agreement were the Aérospatiale SA330E Puma and the Aérospatiale SA341 Gazelle. Puma HC1 XW233 demonstrates its underslung load capability by transporting a Gazelle AH1 for repairs. (Blue Envoy Collection)

Select Bibliography

Aeroplane, various issues, Key Publishing Ltd
AIR 2/14890 General purpose helicopter ASR No OR 325: Bristol 192 (Belvedere)
AIR 2/16585 AST 358: utility helicopter (later, Chinook)
AVIA 18/3468 Comparative flight evaluation of Stallion and Chinook helicopters against ASR358.
AVIA 54/916 Proposed use of Bristol 173 helicopter against Naval Staff Requirement NA.43
Barnes, CH, *Bristol Aircraft Since 1910*, Putnam (1988)
DEFE 48/391 Effectiveness of anti-tank helicopters, National Archives
DEFE 68/80 Helicopters: mounting of anti-tank guided weapons on helicopters, National Archives
DEFE 70/476 Future Army Helicopter requirements, National Archives
Dowling, J, *RAF Helicopters – The First Twenty Years*, HMSO (1992)
Flight International, various issues
Gibson, C, *The Air Staff and the Helicopter*, Blue Envoy Press (2017)
Gibson, C, *The General Staff and the Helicopter*, Blue Envoy Press (2020)
International Air Power Review, various issues, Aerospace Publishing Ltd
Jackson, J, *The Admiralty and the Helicopter*, Blue Envoy Press (2018)
James, DN, *Westland Aircraft Since 1915*, Putnam (1991)
Tagg, AE and Wheeler, RL, *From Sea to Air, Sam Saunders' Legacy*, Crossprint (1989)
Taylor, AH; *Fairey Aircraft Since 1915*, Putnam (1988)
World Air Power Journal, various issues, Aerospace Publishing Ltd

Despite its size, the AW101 Merlin was designed to operate from frigates such as the *Duke*-class. Lynx HMA8 XZ236 is accompanying EH101 Merlin PP5 the first standard Navy Merlin. PP5 conducted trials with HMS *Norfolk* and HMS *Iron Duke* during 1989, being assigned the military serial ZF649. (Blue Envoy Collection)

Glossary

AAC – Army Air Corps
ALAT (Aviation Légère de l'Armée de Terre) – French Army Aviation
Armed Helicopter – Any helicopter with a mounted weapon, e.g. Sioux AH1 with a GPMG
ASM – Air-to-Surface Missile
ASR – Air Staff Requirement
AST – Air Staff Target
ASW – Anti-Submarine Warfare
ATGW – Anti-Tank Guided Weapon
Attack Helicopter – Dedicated anti-armour/support helicopter, e.g. Apache AH1
BERP – British Experimental Rotor Programme
Casevac – Casualty Evacuation
EHI – European Helicopter Industries
FLIR – Forward-Looking Infra-Red
FPB – Fast Patrol Boat
Helarm – Helicopter, Armed – usually applied to British Army helicopters carrying ATGWs
LOH – Light Observation Helicopter (known as a 'Loach' in the US Army)
MAD – Magnetic Anomaly Detector
MATCH – MAnned Torpedo-Carrying Helicopter
MDAP – Mutual Defense Assistance Program
MOD – Ministry of Defence
SAR – Search and Rescue
SARAH – Search and Rescue Homing Beacon
SARBE – Search and Rescue Beacon
TOW – Tube-launched, Optically tracked, Wire-guided
Trigat – Third Generation Anti-Tank weapon